FANTASTIC FACTS ABOUT

THE HUMAN BODY

Created and Written by
John Farndon and Angela Koo for Barnsbury Books

Image Co-ordination
Ian Paulyn

Production Assistant
Rachel Jones

Editorial Director
Paula Borton

Design Director
Clare Sleven

Publishing Director
Jim Miles

This is a Parragon Book
First published in 2000

Parragon, Queen Street House, 4 Queen Street, Bath, BA1 1HE, UK

Parragon has previously printed this material in 1999 as part of the Factfinder series

2 4 6 8 10 9 7 5 3 1

Produced by Miles Kelly Publishing Ltd
Bardfield Centre, Great Bardfield, Essex CM7 4SL

ISBN 0-75253-384-3

Printed in Italy Milanostampa Caleppio Milano

FANTASTIC FACTS ABOUT

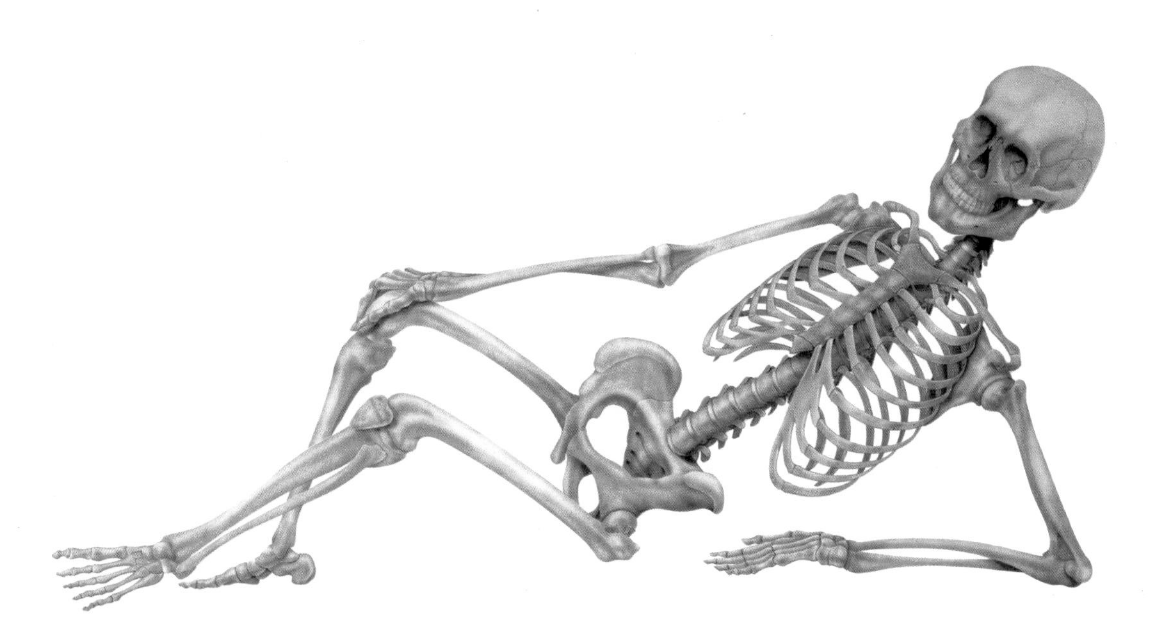

THE HUMAN BODY

p

CONTENTS

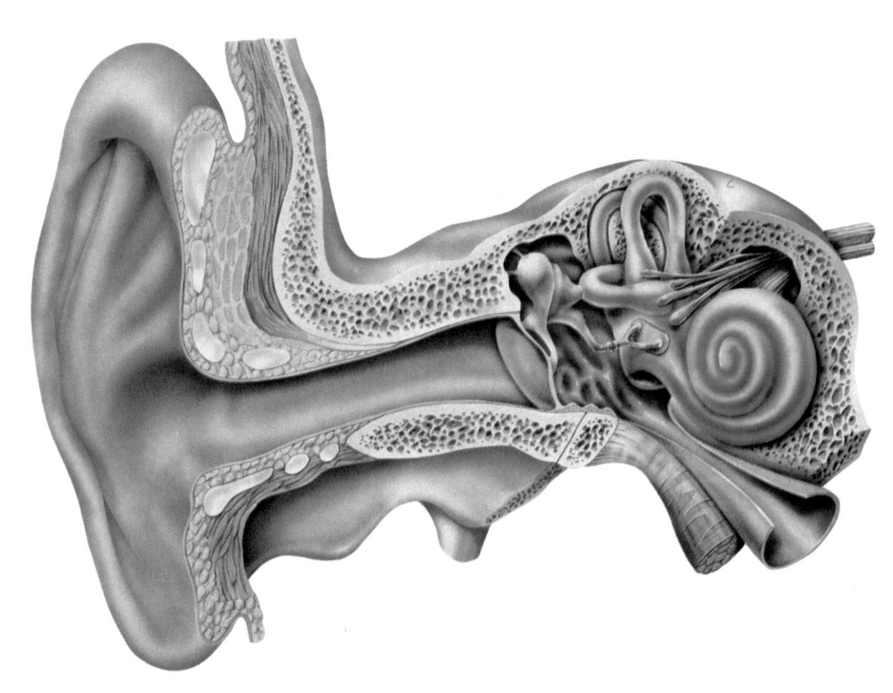

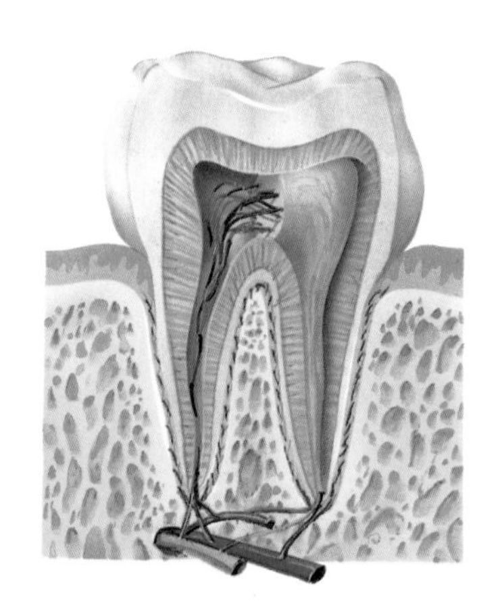

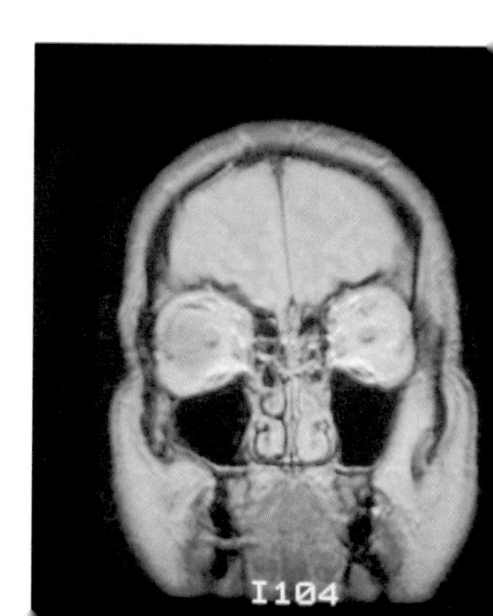

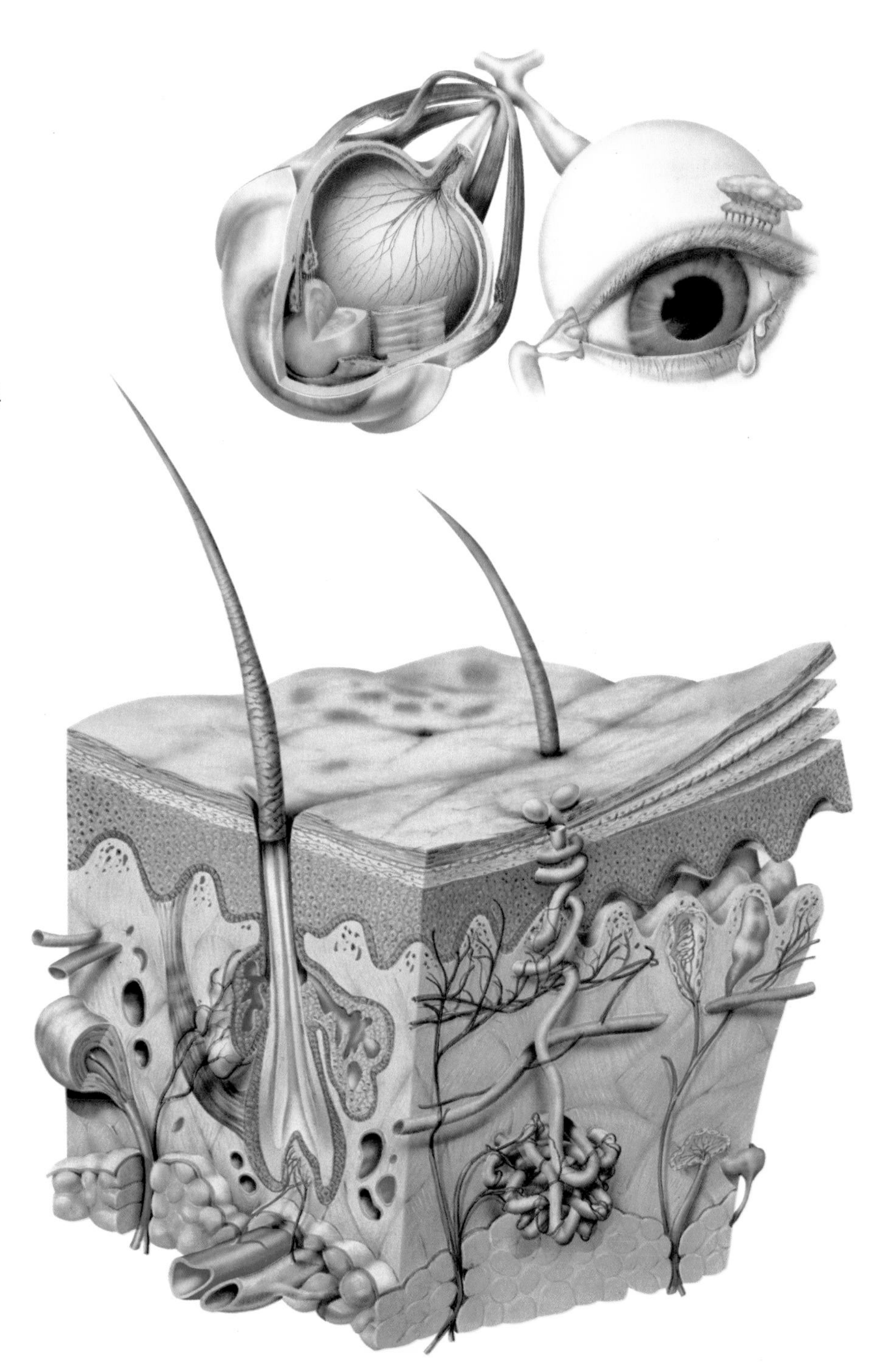

INTRODUCTION

There is perhaps nothing more mysterious or remarkable in nature than the human body. Find out what the body is made of, from the simplest cells to the astonishing complexity of the human brain. Discover how the body works through its organs, muscles and bones and its amazing systems of nerves and arteries. Finally, discover the amazing facts governing human reproduction and the miracle of new life.

HUMAN BODY is a handy reference guide in the *Fascinating Facts* series. Each book has been specially compiled with a collection of stunning illustrations and photographs which bring the subject to life. Hundreds of facts and figures are presented in a variety of interesting ways and side-panels provide information at-a-glance. This unique combination is fun and easy to use and makes learning a pleasure.

BODY CELLS

Cells are the basic building blocks of our bodies. Just as a house is built of bricks, so bodies are built of microscopic cells. There are 60 million or so in the body, each with its own special task. Some are skin cells. Some are liver cells. Some are fat cells. Each system in the body has its own special cells.

Body cells come in many shapes and sizes. But they are all squidgy cases of chemicals. Holding them together is a thin casing of fat dotted with protein called the membrane. Although it holds the cell together, it lets certain chemicals move in and out. Inside the cell is watery fluid called cytoplasm, and floating inside the cytoplasm is the cell's nucleus — the control centre which contains the chemical instructions for all the cell's tasks. Every time a new chemical is needed, the nucleus sends the instructions to the rest of the cell.

LIVING CHEMICAL FACTORY
Every cell in your body is a bustling chemical factory. Inside, every second of the day, the cell's team of 'organelles' is ferrying chemicals to and fro, breaking up unwanted chemicals, making new ones, using them and sending them off to other cells. Instructions come from the nucleus, but every bit of the cell has its own allotted task.

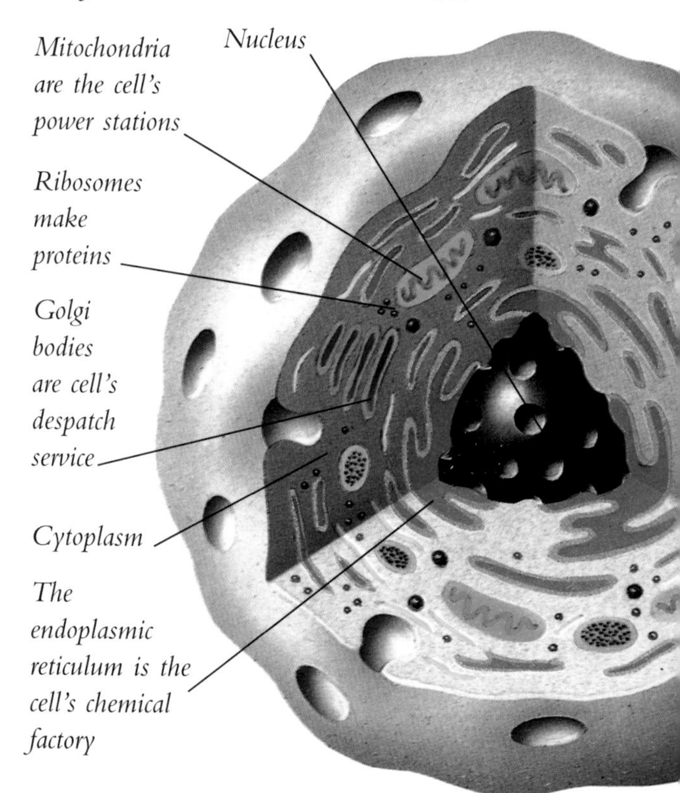

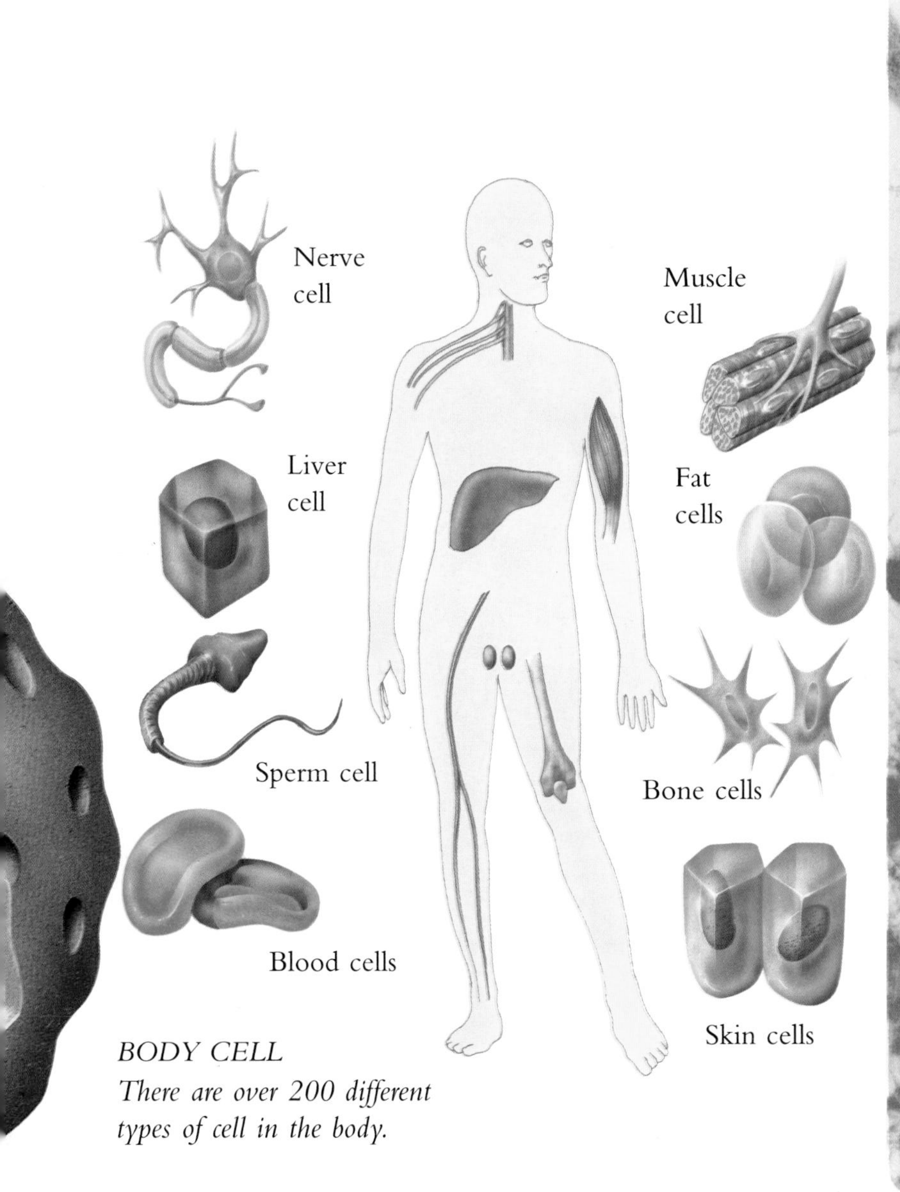

BODY CELL
There are over 200 different types of cell in the body.

Nucleus and organelles

- The nucleus is the cell's control centre, housing the remarkable spiral-shaped chemical molecule DNA. In a chemical code, DNA not only gives all the instructions the cell ever needs to carry out its task, but also carries a complete blueprint for another you.
- Outside this is a mixture of dissolved chemicals and floating structures called organelles. Each organelle has a particular function.
- Mitochondria are the cell's power houses, turning chemical fuel supplied by the blood into energy packs of the chemical ATP.
- The intricate layers of the Golgi bodies are the cell's despatch centre, where chemicals are stowed in tiny membrane bags.
- Ribosomes make proteins from basic chemicals called amino acids.
- Lysosomes are the waste disposal units, breaking up unwanted material.

TISSUES

All the body's many kinds of cells group together to make substances called tissues. Each tissue is made almost entirely of particular kinds of cells packed together, and each has its own purpose. Nerve tissue, for instance, is made of identical cells called neurons, which are good at sending electrical signals. Epithelial tissue, which makes skin and other tissues, is made of three kinds of cell which make a waterproof layer. Typically, the spaces between tissue cells are filled with a fluid called tissue fluid.

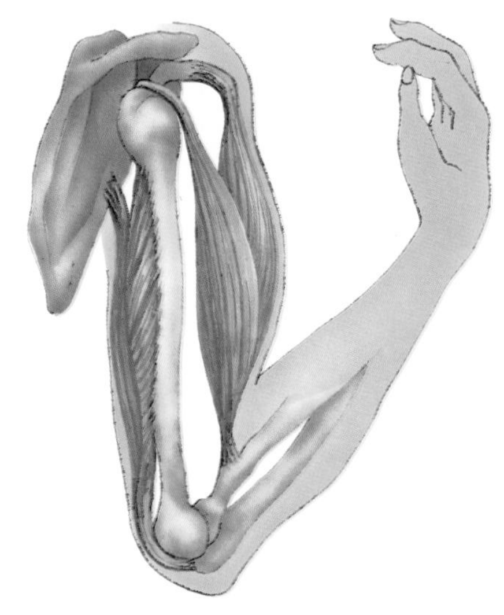

MUSCLE TISSUE

Muscle tissue makes up the bulk of your body — almost half its entire weight. It is tissue built of special long, usually reddish cells that are able to pull shorter and then relax.

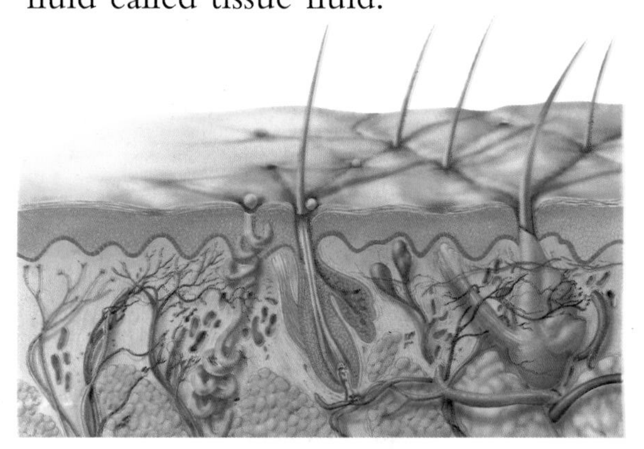

EPITHELIAL TISSUE

Epithelial tissue is lining tissue, built up from three basic cell shapes — squamous (flat), cuboid (box-like) and columnar (rod-like). Most is in layers just one cell thick — lining inside blood vessels, airways, inside the heart and chest and many other places. But skin is made of many layers.

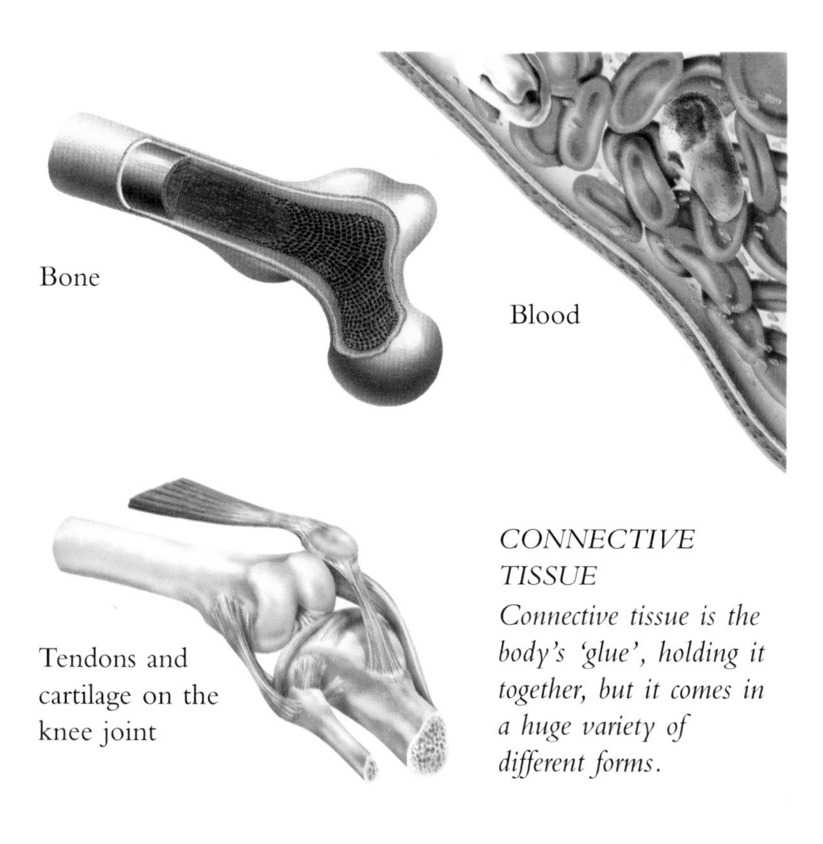

CONNECTIVE TISSUE

Connective tissue is the body's 'glue', holding it together, but it comes in a huge variety of different forms.

CONNECTIVE TISSUE

'Connective' tissue forms in the space between other tissues and helps hold the body together. But it comes in a variety of different forms including 'adipose tissue', better known as fat, tendons and cartilage. Bone and even blood are connective tissues. Of course, blood does not actually connect anything, but it begins life in the human embryo in the same way.

Tissues

Organs

- Different types of tissue group together to make organs.
- The bulk of your heart is made of muscle tissue. But it is lined with epithelial tissue, nerve tissue sends signals to it, and it is filled with the connective tissue of blood.

Connective tissue

- Connective tissue is made of three main things: cells, fibres and 'matrix'.
- Cells are mainly 'fibroblasts', cells which make fibres. But they also include fat cells and 'macrophages', which eat germs.
- The fibres are made from microscopic strings of protein such as white ropey collagen, stretchy elastin, and branch-like reticulin.
- The matrix is the basic setting for the other materials, like the bread in a currant loaf. It can be anything from a runny syrup to a thick gel.

BODY SYSTEMS

Although your body seems very complicated, it all makes sense if you think of it in terms of a number of systems, each with its own tasks to do. It only seems complicated because these systems are all interlinked.

Some of these systems extend throughout the body, like the skeleton which is the body's framework, the musculature which is the body's means of moving, and the nervous system which is the body's communication networks. Others are

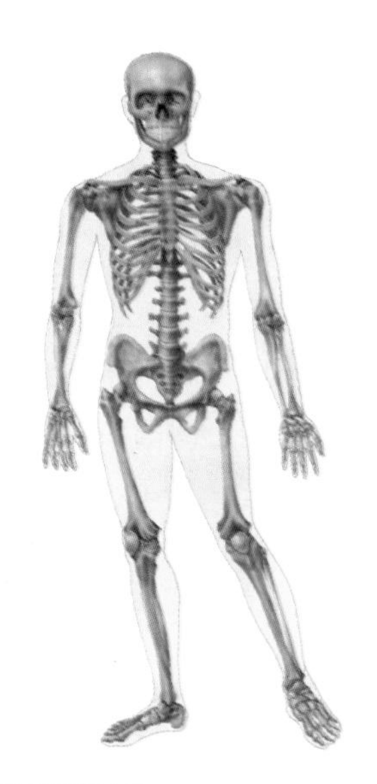

SKELETON
The skeleton supports the body and protects internal organs such as the lungs and the heart.

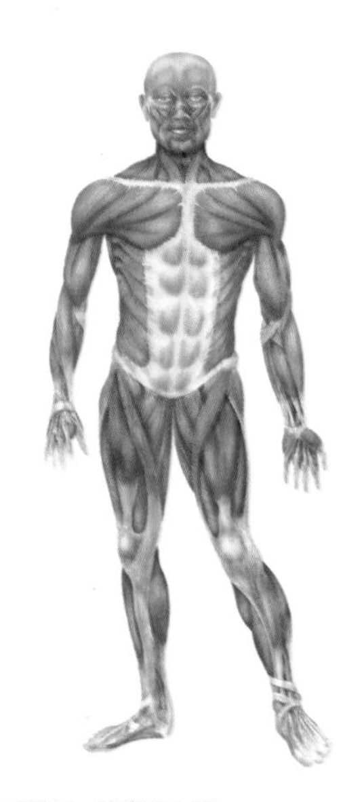

MUSCULATURE
The muscles are what enable you to move. Muscles are also involved in other body systems.

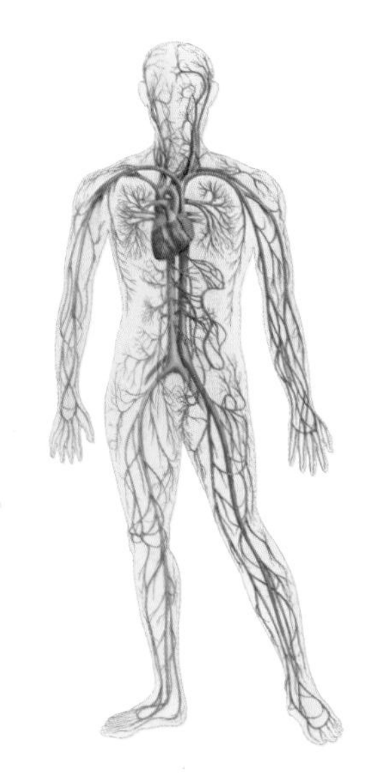

CARDIOVASCULAR SYSTEM
Heart and blood circulation supply body cells with oxygen and food, and defend the body against germs.

NERVOUS SYSTEM
The nervous system is the brain and the nerves — the body's high speed control network.

quite localized, like the digestive system, which is the body's food processor, and the urinary system, which controls the body's water. The body could not function without any of these systems, and you go on living because they all work together to keep you alive.

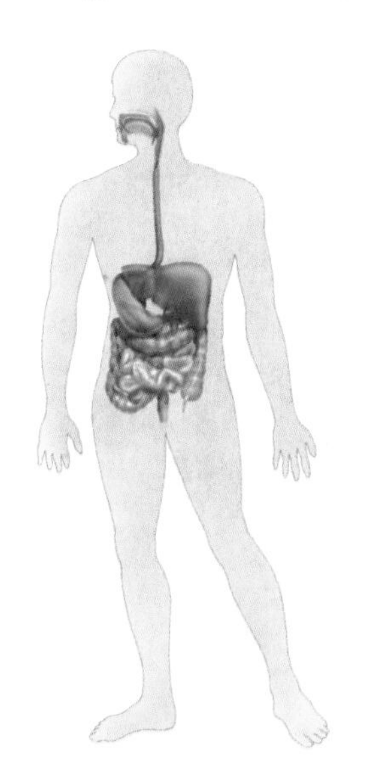

DIGESTIVE SYSTEM
The digestive system breaks down food and turns it into the right chemicals for the body to use.

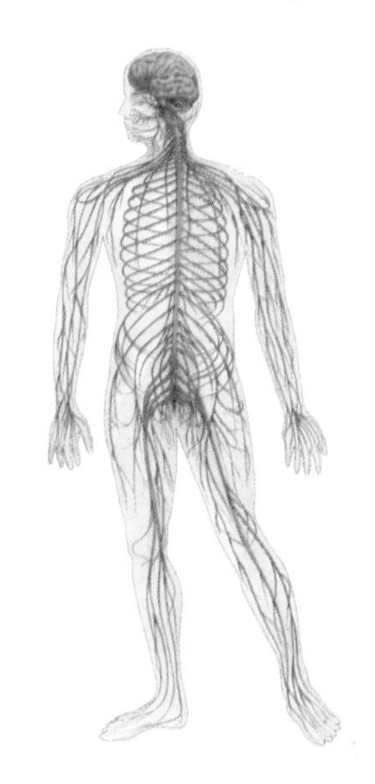

LYMPHATIC SYSTEM
Lymph fluid containing immune cells that help fight disease circulate in lymphatic ducts.

Body systems

- Skeleton, or skeletal system.
- Musculature, or muscle system.
- Cardiovascular system.
- Nervous system.
- Digestive system.
- Excretory system.
- Urinary system: controls the body's water and removes excess as urine.
- Immune system, including the lymphatic system, white blood cells and antibodies: the body's various defences against germs.
- Respiratory system: takes air in and out of your lungs through the mouth and nose, giving oxygen to the blood and taking out carbon dioxide.
- Reproductive system: the smallest of the systems, this is essentially the genitals and the organs connected to them — the organs that enable us to have children. This is also the only body system that can be surgically removed without threatening your life.

BREATHING

You have to breathe to stay alive. When you breathe in, you take air into the body — and air contains the oxygen vital to every cell in your body. Just as a fire needs air to burn, so body cells need oxygen to burn up the food they get in the blood. Without it, they die — which is why if you stopped breathing for even a few minutes, you would lose consciousness, and eventually die. So, several times a minute your body sucks air into your lungs so that oxygen can be spread around the body in the blood.

BREATHING AT ALTITUDE
The air high on a mountain contains less oxygen, so mountaineers must breathe and circulate blood faster to compensate — or carry a supply of oxygen.

PANTING HARD
The cells in a runner's muscles are working overtime, burning up sugar at a huge rate to keep him pounding along. Burning all this sugar means the cells need lots of oxygen, so he must breathe hard and his heart must pump blood fast to supply the muscles with all the oxygen they need. As a runner trains and gets fit, his body gets much quicker at raising oxygen levels. The unfit are left behind, gasping for breath.

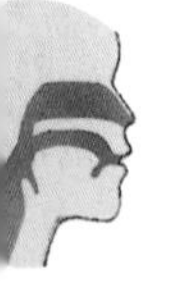

WHY BREATHE OUT?

Food arrives at each body cell as a chemical called glucose which is made mainly of carbon and hydrogen. When the cell burns glucose to release its store of energy, the hydrogen joins with oxygen to make water and the carbon joins with oxygen to make carbon dioxide. Carbon dioxide is poisonous to the body and must be removed. This is what happens when you breathe out.

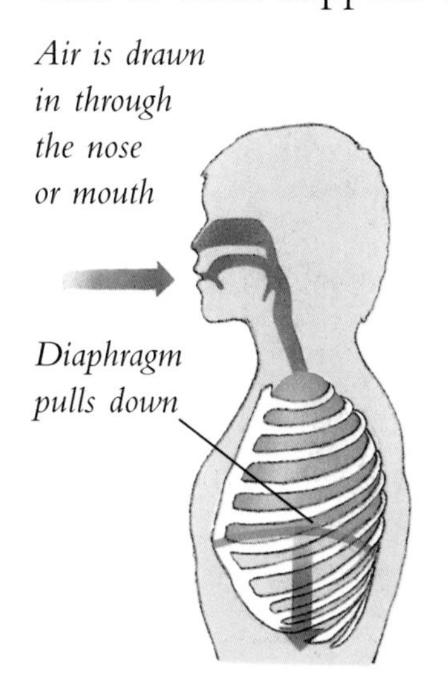

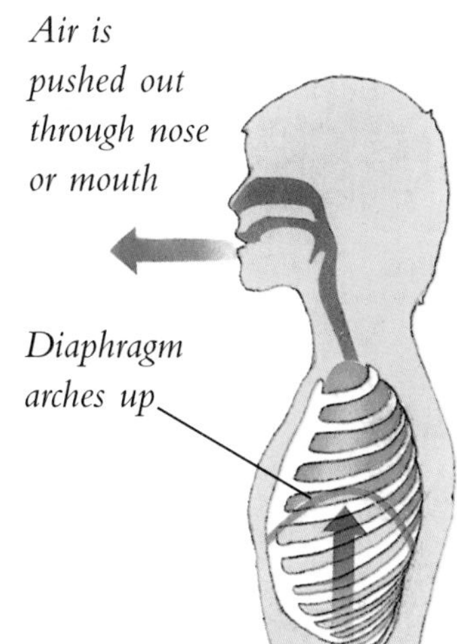

BREATHING IN

To draw air in, the lungs must expand. 'Intercostal' muscles around your ribs pull your chest up and out, and a domed sheet of muscle called the diaphragm pulls the lungs down.

BREATHING OUT

Lungs are spongey and elastic, so collapse like a deflated balloon as you breathe out — but can only collapse so far because a layer of 'pleural fluid' sticks them to the chest walls.

Breathing

- You will probably take about 600 million breaths if you live to the age of 75.
- Every minute you breathe, you take in about 6 litres of air.
- A normal breath takes in about 0.4 litres of air. A deep breath can take up to 4 litres.
- On average, you breathe about 13-17 times a minute. But if you run hard, you may have to breathe up to 80 times a minute.
- Newborn babies breathe about 40 times a minute.
- Breathing is called respiration. The way cells burn up glucose using oxygen is called cellular respiration.
- Air normally contains about 21% oxygen, and about 0.03% carbon dioxide. The air you breathe out has about 0.6% carbon dioxide — not quite enough to poison anyone with your breath!

THE LUNGS

Your lungs are a pair of soft bags in your chest. Like spongey cushions, they are riddled with tiny branching tubes which join together to lead right up to your mouth. Whenever you breathe in, air is sucked in through your mouth or nose and rushes down your windpipe or trachea until it reaches a fork deep inside your chest. At the fork, the airways branch into two, one branch or bronchus leading to the left lung and the other to the right.

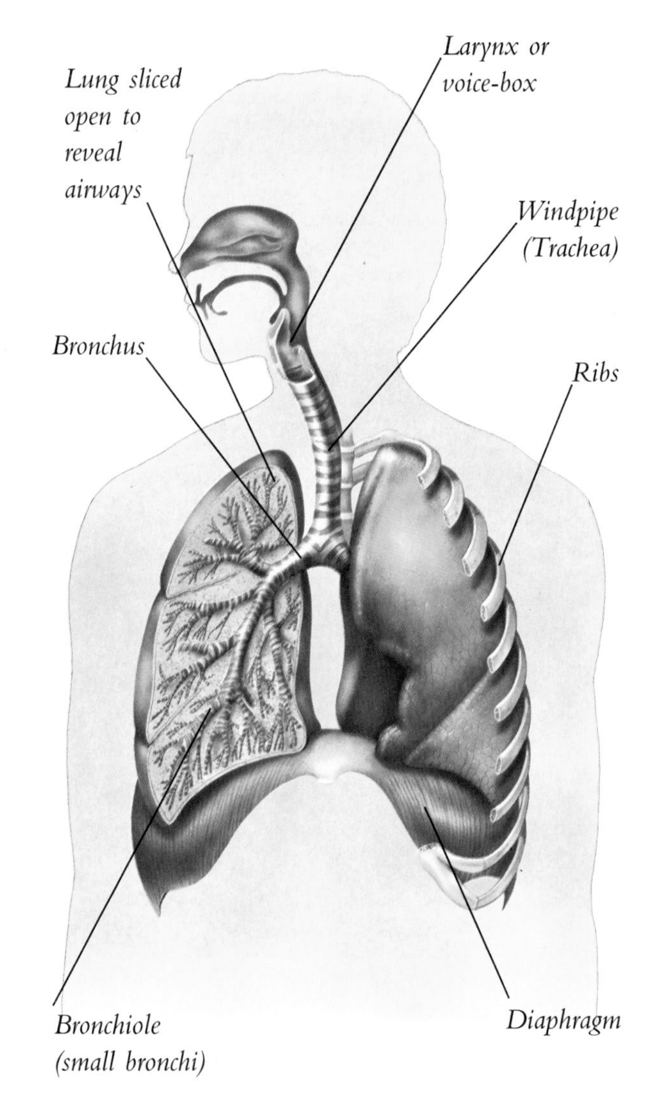

INSIDE THE LUNGS

Inside the lungs, the bronchi branch into millions of bronchioles. There are so many tiny airways in the lungs that they have a gigantic surface area. It is this huge surface area that allows a great deal of oxygen to seep through into the surrounding blood vessels in just a few seconds every time you breathe.

ALVEOLUS
Alveoli (right) are the tiny air sacs at the end of each airway that fill up like balloons as you breathe in. Here oxygen passes through into the blood.

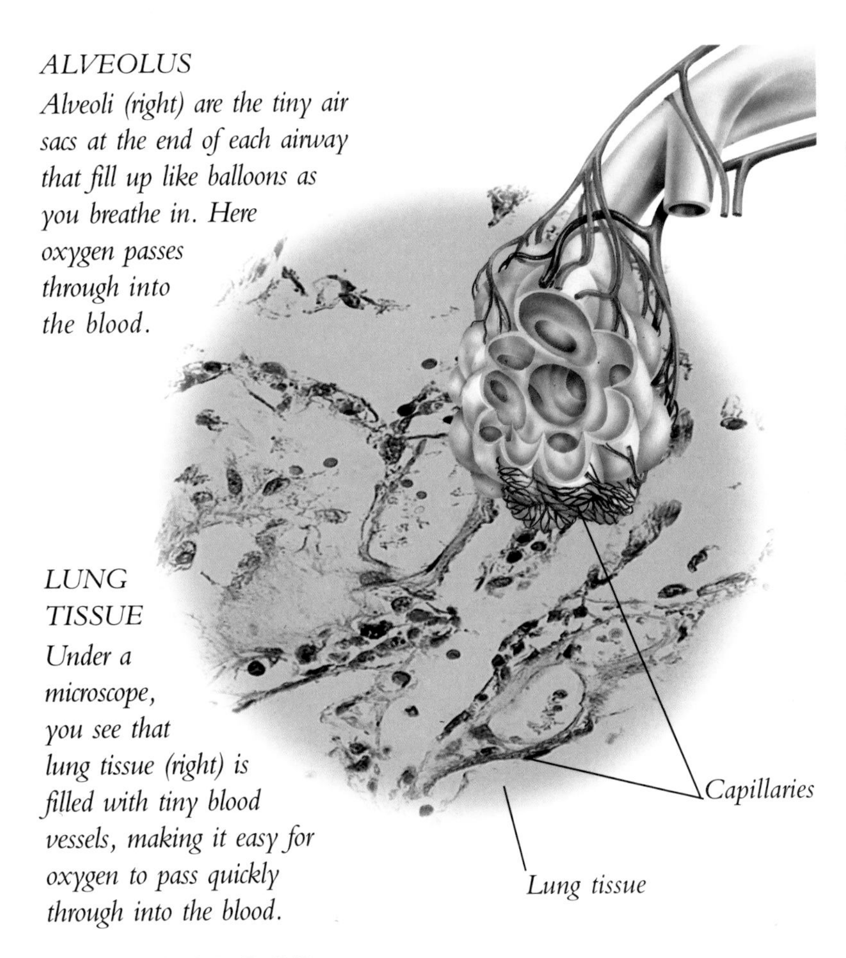

LUNG TISSUE
Under a microscope, you see that lung tissue (right) is filled with tiny blood vessels, making it easy for oxygen to pass quickly through into the blood.

AIR TO BLOOD

It is in the lungs that oxygen in the air is absorbed into the blood. Around the end of each bronchiole are bunches of tiny air sacs called alveoli. The walls of each sac are just one cell thick, and here oxygen seeps through into the minute blood vessels wrapped around them.

Lungs

The lung surface

- There are about 300 million alveoli in your lungs.
- Open out and laid flat, the inside of the alveoli would cover an area the size of a tennis court.
- There are over 2,400km of airways in your lung.

Coughing and hiccupping

- The surfaces of the airways are protected by a film of slimy liquid called mucus. When you've a cold or a chest infection, the airways may fill up with mucus, making you cough to clear them.
- Smoking irritates the airways and makes them fill up with mucus. It also weakens the tiny hairs or 'cilia' that waft the mucus out. So lungs are prone to infection. Smoking also increases the risk of lung cancer.
- Hiccups are caused by a sudden contraction of the diaphragm, dragging air into your lungs so quickly that your vocal cords snap shut.

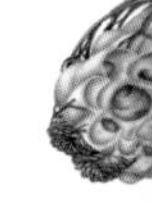

CIRCULATION

Once in the blood, oxygen must be delivered swiftly to the cells where it is needed. At the same time, unwanted carbon dioxide must be picked up from the cells and brought back to the lungs for breathing out. This is what the blood circulation is for.

Driven by the heart, blood is pumped through an intricate network of blood vessels all the way round the body again and again, once every 90 seconds. As it circulates it goes through the lungs where it washes around the lung's millions of air sacs and picks up oxygen. This bright red, oxygen-rich blood then goes on back through the heart, where the heart's powerful right side gives an extra boost to

BLOOD VESSELS

The blood circulates through the body in a series of pipes called blood vessels. Oxygen-rich blood is bright red and travels through arteries on the way out from the heart. Oxygen-poor blood is purple and travels in veins on its way back to the heart.

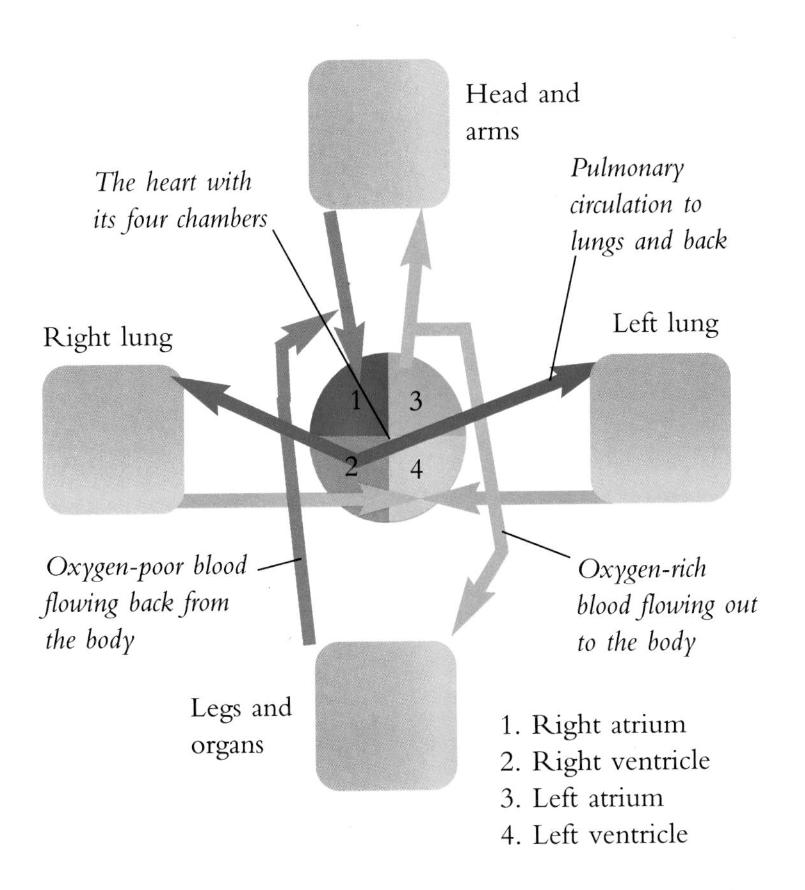

push it into the arteries and right round the body. As it passes into ever narrower blood vessels — first arterioles, then capillaries — it delivers its oxygen to the cells and picks up the waste carbon dioxide. It then flows back through the veins to the left side of the heart — now much bluer in colour because it is short of oxygen — ready to be pumped on to the lungs for new intake of oxygen.

Circulation

- The blood circulation has two parts: the systemic and the pulmonary.
- The pulmonary circulation is the short section that sends oxygen-poor blood from the heart to the lungs and brings it back to the heart with fresh oxygen.
- The systemic circulation sends oxygen-rich blood on from the right side of the heart and carries it right round the body and back to the left side of the heart.
- Oxygen is ferried through the blood in red blood cells.
- Each of the millions of red blood cells can carry plenty of oxygen because they contain a substance called haemoglobin, which takes up oxygen when there is plenty around, and gives it up when there is a shortage.
- Haemoglobin glows bright scarlet when it is carrying oxygen, which is why blood is red. But it fades to dull purple when it loses oxygen.

THE HEART

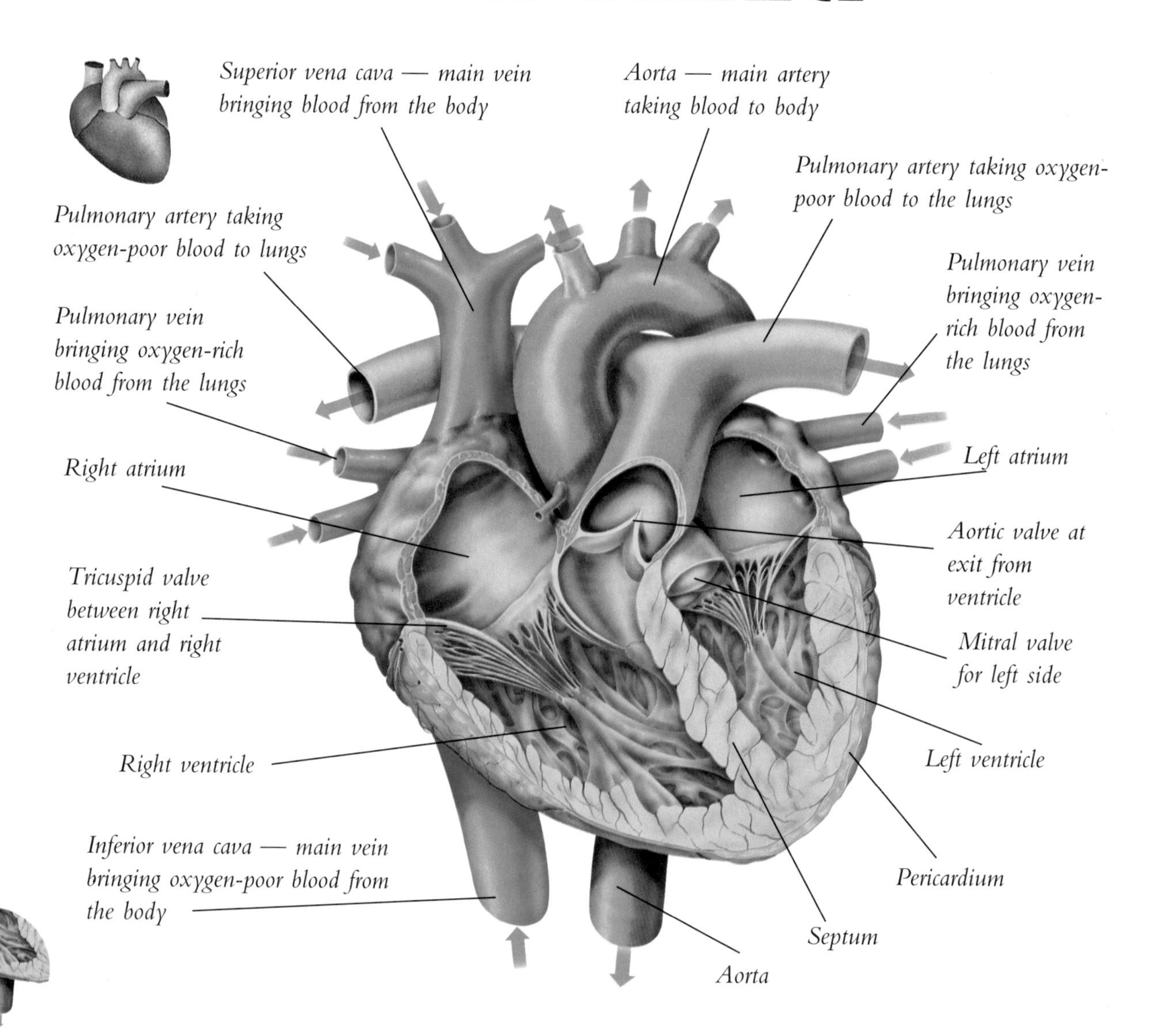
Superior vena cava — main vein bringing blood from the body
Aorta — main artery taking blood to body
Pulmonary artery taking oxygen-poor blood to the lungs
Pulmonary artery taking oxygen-poor blood to lungs
Pulmonary vein bringing oxygen-rich blood from the lungs
Pulmonary vein bringing oxygen-rich blood from the lungs
Right atrium
Left atrium
Aortic valve at exit from ventricle
Tricuspid valve between right atrium and right ventricle
Mitral valve for left side
Left ventricle
Right ventricle
Inferior vena cava — main vein bringing oxygen-poor blood from the body
Pericardium
Septum
Aorta

The heart is a remarkable little pump, about the size of a fist and made of pure muscle. Every second of your life, the heart muscle is squeezing away, pumping blood around your body in a continuous stream. About 70 times a minute — faster if you're running about — the muscles contract to send jets of blood shooting through the arteries, the big pipes that lead away from the heart. Press a hand against the middle of your chest, just to the left, and you can often feel the heart pounding away inside.

The heart owes its steady beat to the special muscle it is made from. Most muscle has to be prodded into action by nerve signals. Heart muscle contracts and relaxes rhythmically by itself. If it has blood to keep it going, it can go on working even outside the body. Outside stimuli can only change its rate.

INSIDE THE HEART

The heart is not just one pump but two, each forming one side of the heart. A thick wall of muscle called the septum keeps the sides completely separate. The left side is the stronger pump, driving oxygen-rich blood from the lungs all around the body. The right is the weaker pump, pushing it only through the lungs to pick up oxygen. Each side has two chambers separated by a one-way valve — an 'atrium' at the top, where blood accumulates and a 'ventricle' below, which is the main pumping chamber.

Heart

- The heart beats more than 30 million times a year.
- All the chambers of the heart hold about the same quantity of blood — about 70-80ml.
- The heart has four valves to ensure blood flows only one way, two on each side of the heart — a large valve between the atrium and ventricle and a smaller one at the exit to the ventricle.
- The large valves have different names on each side of the heart. The valve in the left is called the mitral valve; the valve in the right is called the tricuspid valve.
- Heart muscle gets its blood from the blood vessels supplied by the coronary arteries, which wrap around the outside. If these arteries get blocked, the heart muscle may be deprived of blood and stop working. This is what happens in a heart attack.

ARTERIES AND VEINS

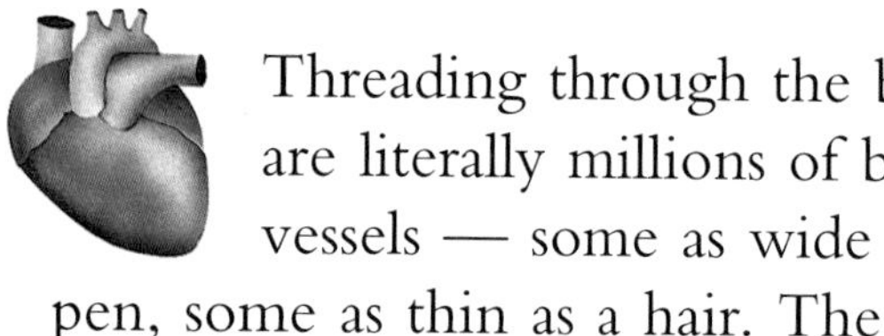

Threading through the body are literally millions of blood vessels — some as wide as a pen, some as thin as a hair. The biggest blood vessels carrying blood away from the heart are the arteries. These branch into arterioles, and these in turn branch into capillaries. From the capillaries, blood flows into wider venules, and then into wider still veins on its way back to the heart. Blood vessels are not just passive pipes. All but capillaries have muscles and valves to control the way blood flows, helping smooth out the flow.

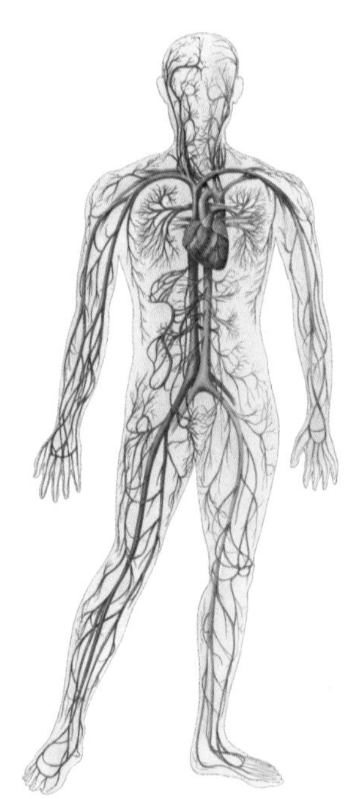

BRANCHING NETWORK
The body's network of blood vessels is like two river systems intertwined. This illustration shows only the arteries and arterioles in red in one half of the body and only the veins and venules in blue in the other. Of course, both veins and arteries supply both sides of the body, and the networks are duplicated.

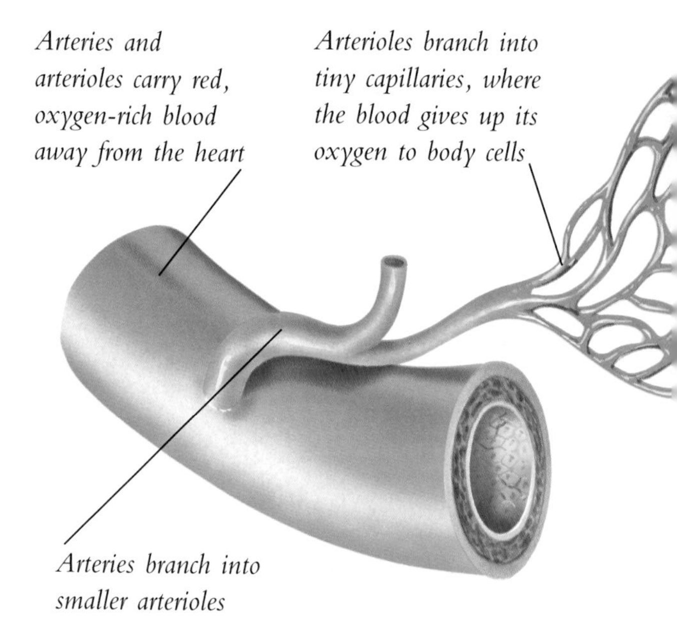

Arteries and arterioles carry red, oxygen-rich blood away from the heart

Arterioles branch into tiny capillaries, where the blood gives up its oxygen to body cells

Arteries branch into smaller arterioles

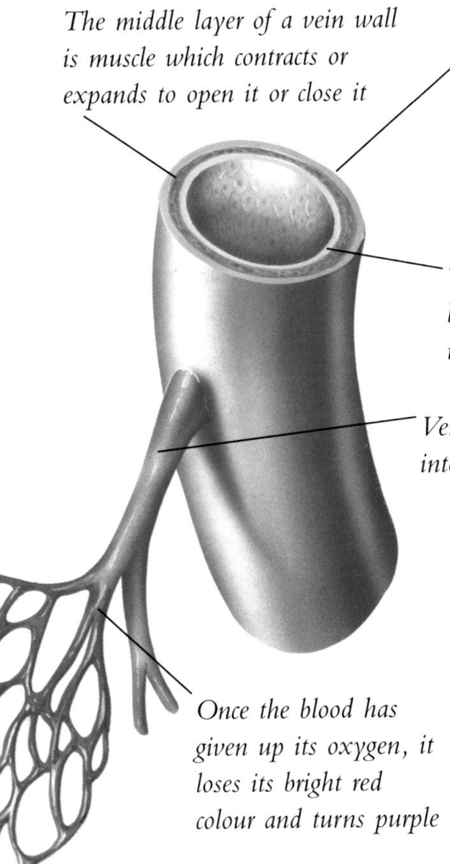

BLOOD VESSELS
Arteries and veins are the blood circulation's main highways. They are linked by small arterioles and venules and tiny capillaries.

DYNAMIC PIPES

The muscular walls of the blood vessels control the flow of blood, narrowing or widening them, for instance, to divert blood to where it is needed. When tissues such as muscles are active, blood vessels to them open up to increase blood supply. When tissues are resting, some of the blood vessels close.

Blood vessels

- Blood races through the arteries at 1 metre a second.
- There are over 60,000km of capillary in the body.
- Many capillaries are a hundredth of a millimetre thick.
- The muscles of the blood vessel walls control the pressure of blood flow. They tighten and relax so that no matter how much blood is delivered by the main arteries, pressure in the capillaries is always right. It must be strong enough to push oxygen to every cell, but not so strong as to burst the capillaries.
- High blood pressure is called hypertension. It is caused by the thickening of the artery walls, which makes the arteries narrow.
- Blood pressure is lower in veins than it is in arteries.
- The walls of veins are thinner, weaker and less elastic than those of arteries. They even collapse when empty, unlike arteries.

Blood

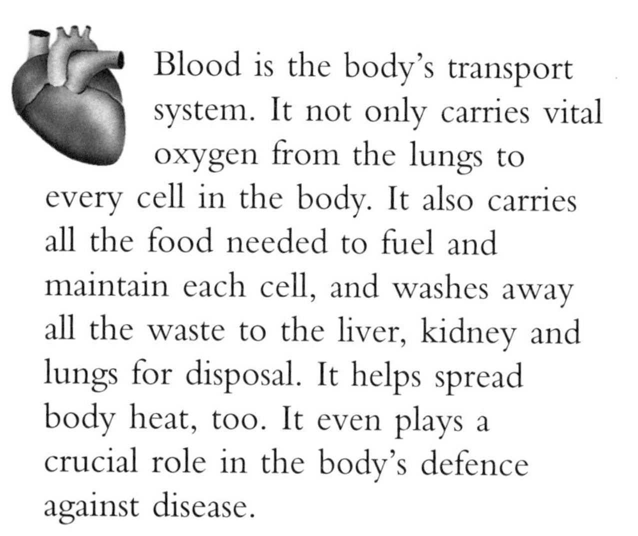

Blood is the body's transport system. It not only carries vital oxygen from the lungs to every cell in the body. It also carries all the food needed to fuel and maintain each cell, and washes away all the waste to the liver, kidney and lungs for disposal. It helps spread body heat, too. It even plays a crucial role in the body's defence against disease.

WHAT'S BLOOD MADE OF?

Blood looks like red ink, but under a microscope, you can see a rich variety of ingredients, swept along in a clear, yellowish fluid called plasma. Besides many chemicals, there are three main kinds of cell. Most numerous are the button-shaped red blood cells that carry oxygen. Then there are tiny lumps called platelets. Finally there are giant white cells or 'leucocytes'.

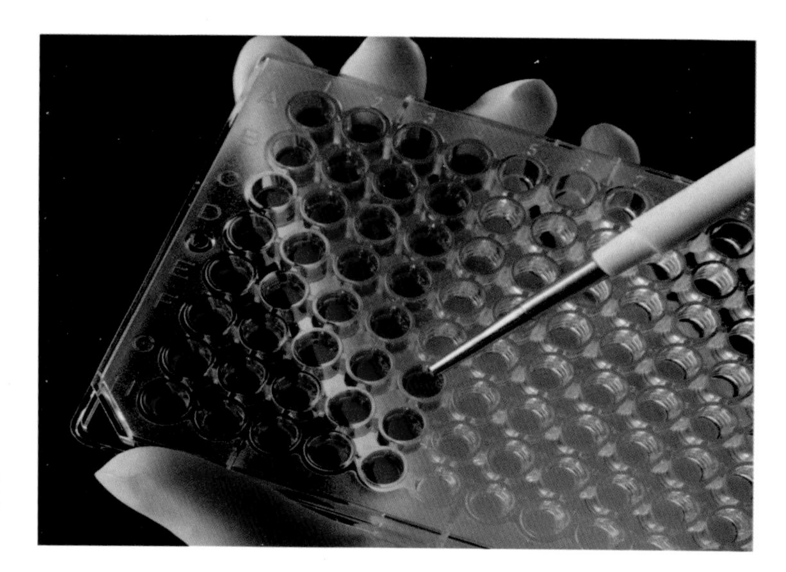

BLOOD SAMPLE

Blood need only be taken in very small quantities to reveal a great deal about it. There are, for instance, five million red blood cells in every cubic millimetre of blood. Red cells, white cells and platelets make up half the blood in any sample. The rest is the fluid plasma, which includes a solution of the chemical albumin, antibodies and 'clotting factors' — chemicals which make the blood clot (thicken to stop bleeding).

DROP OF BLOOD

Under a microscope, you can see some of the many different ingredients in blood. It is dominated by the red blood cells which turn bright red only when they are carrying oxygen. This gives blood its red colour. Otherwise, blood is purply brown.

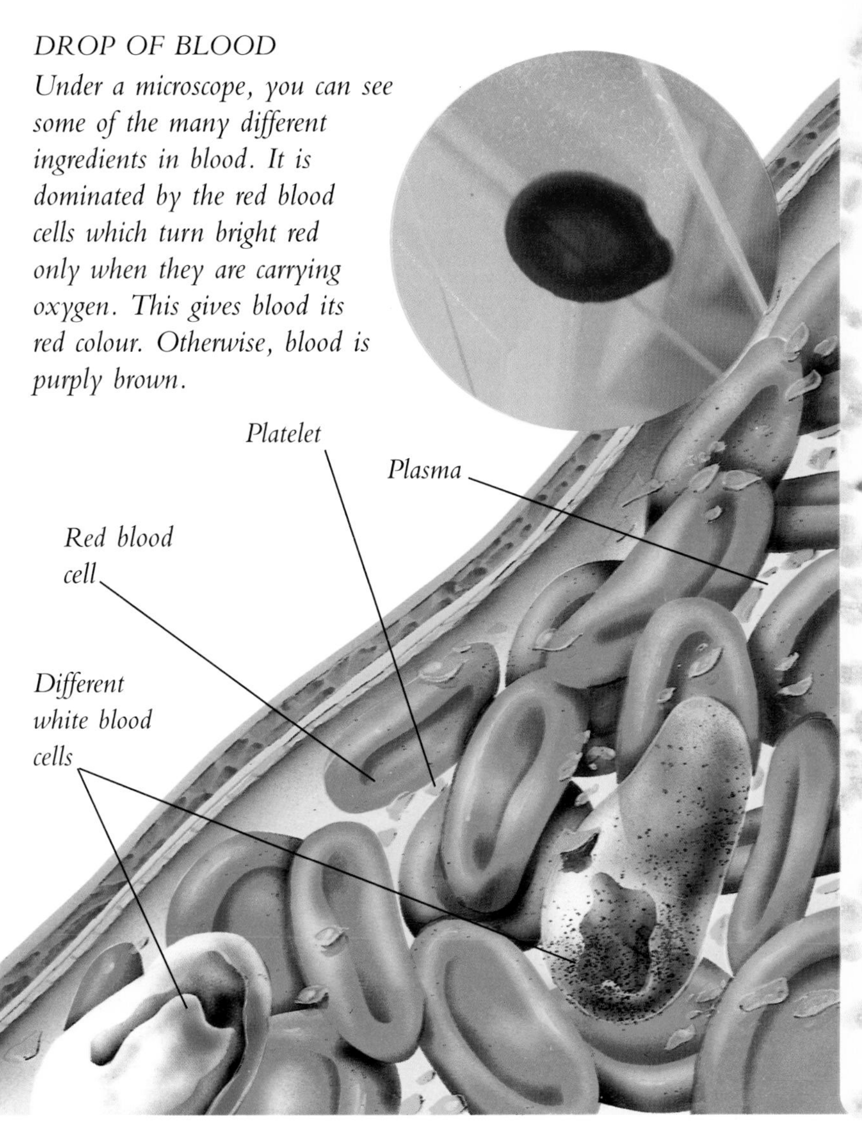

Blood cells

Red blood cells

- Red blood cells are the button-shaped cells which contain haemoglobin, the special chemical that holds oxygen as it is ferried by the cell around the body.

White blood cells

- White cells are involved in the body's defence.
- Most white cells contain tiny grains and are called granular leucocytes or granulocytes.
- Most granulocytes are called neutrophils, which act as scavengers, eating up intruders.
- There are two other kinds of granulocyte: eosinophils and basophils.

Platelets

- Platelets look like tiny chips and are bits chipped off other cells. They help to plug leaks such as cuts.
- Platelets also slow blood loss by releasing 'clotting activators' which help fibres grow around the wound and stop blood leaking.

SKIN

Your skin is not just a bag for your body. It is actually the body's largest organ and has a wide range of vital tasks. It is a coat that protects your body from the weather and from infection. It helps keep your body at just the right temperature by insulating it from the cold and letting out heat when it is warm. It helps you sense the world around you by responding to touch. It even helps nourish you, by using sunlight to make vitamin D. It is so important that it receives almost a third of your body's blood supply and has a range of special glands.

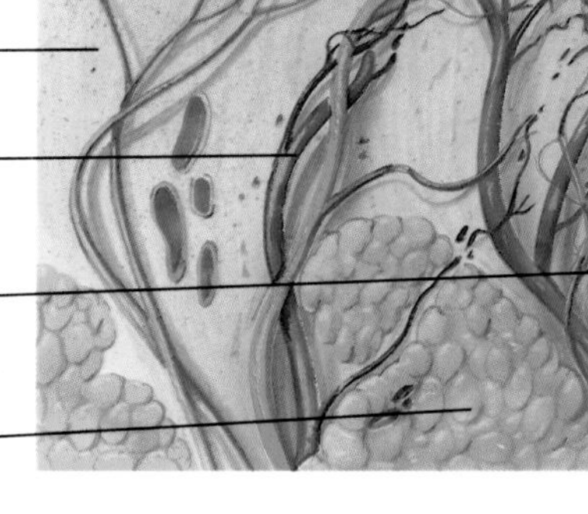

SLICE OF SKIN
This is a hugely magnified view of a slice through the skin, showing just some of the elements. Its thickness varies over the body. It is thickest — about 6mm — on the soles of your feet, and thinnest on your eyelids, where it is just 0.5mm thick.

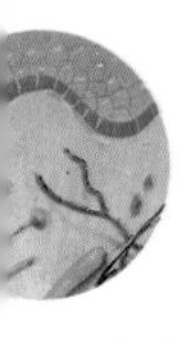

DOUBLE LAYER

Over most of your body, your skin is just 2mm or so thick, and is easily penetrated by a sharp edge. But a microscope shows it is made of two main layers: a thin, essentially dead, outer layer called the epidermis and a thicker lower layer called the dermis, which contains the glands.

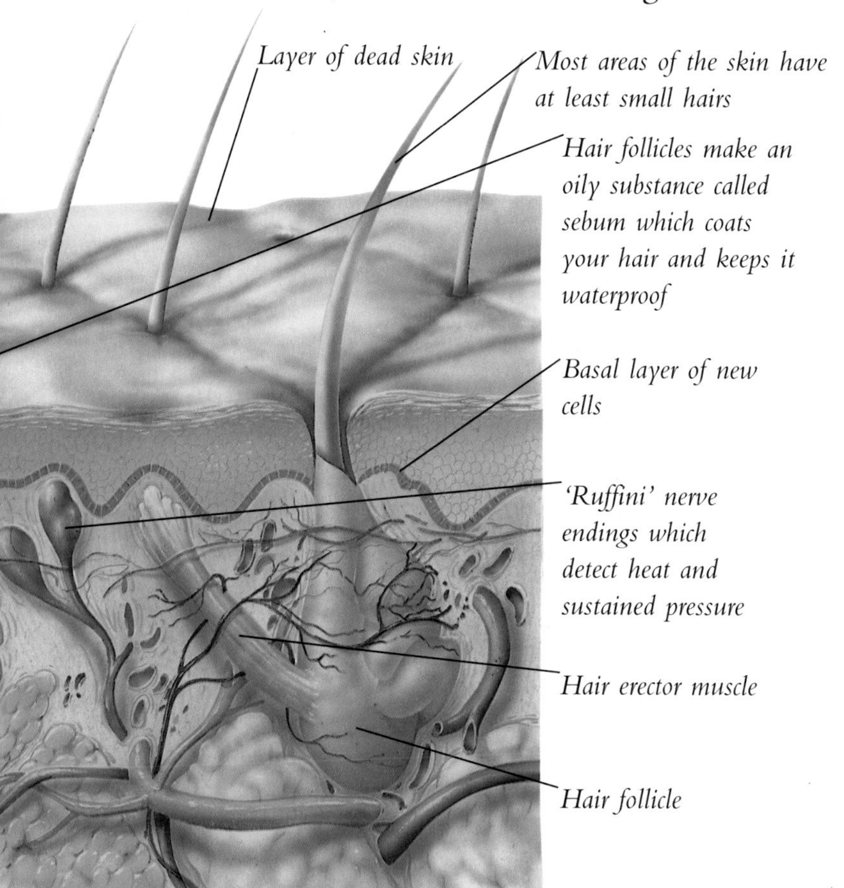

Skin

The epidermis

- New cells are continually growing in the basal layer at the bottom of the epidermis.
- As new cells grow in the epidermis's basal layer, they push old cells up. As they move further away from the blood supply, they die off, leaving just a hard protein called keratin. The layer of keratin gives the skin a hard, protective coat.
- Dead cells continually flake off the skin's surface, but new cells come up to take their place.

Skin colour

- The epidermis contains special cells which produce a dark pigment called melanin. In fair-skinned people, extra melanin is made to protect the skin in strong sunlight, which is why they tan. Dark-skinned people start with much more melanin in their skin. Freckles are just spots of melanin-rich skin.

HAIR AND NAILS

Almost alone among mammals (except for those that live in the sea), we humans have bare skin. Apart from the hair on your head (and around your genitals when you're an adult), your skin has only a light covering of tiny down hairs.

People sometimes talk about dull and lifeless hair when it is in poor condition. In fact, hair is always lifeless, for it is made of keratin which is the dry material left behind by dead cells — the same dead material that nails and the outer layer of skin is made from.

The root of each hair is embedded in the skin, in a pit called the hair follicle. The hair grows as dead cells pile up within this follicle, held in place by the hair's club-shaped end or 'bulb'. Each hair has a spongey core, surrounded by a ring of long fibres which are wrapped around in turn by overlapping layers.

DARK HAIR

Providing you don't dye it, the colour of your hair depends on just how much of a pigment called melanin you have in the shafts of your hairs. This pigment is made in cells called melanocytes near the hair root.

HAIR BASE

A finger nail takes about six months to grow from base to tip, though it varies with the seasons. Like hair, it is made of keratin and grows from a root, called the nail bed. At the base of each nail is a half-moon, called the lunula, covered by a flap of skin called the cuticle.

Lunula

Nail bed

Cuticle

HAIR BASE

Hair is made from dead skin cells and keratin and grows in follicles deep in the skin. At the root, there is an erector muscle which pulls it upright when you are cold. There is also a sebaceous (oil) gland which oozes oil to keep the hair waterproof and stop it from drying out.

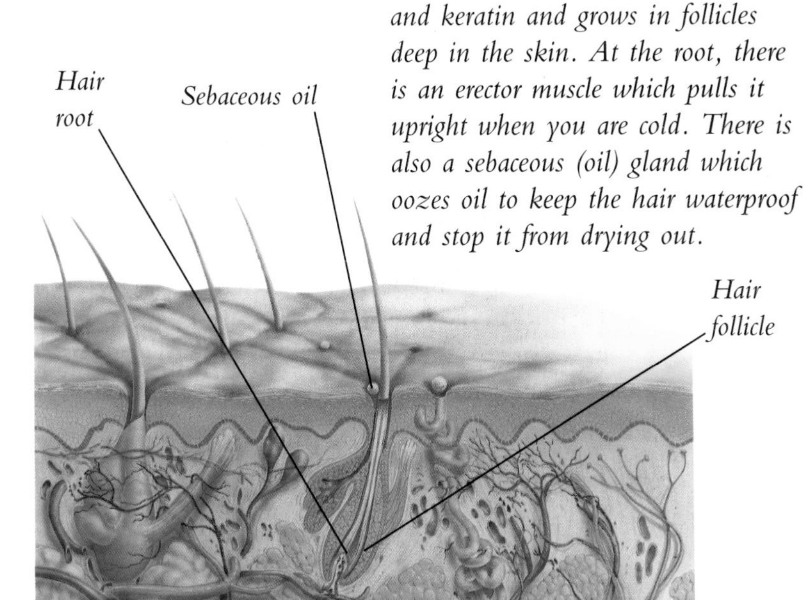

Hair facts

- There are three types of hair on your body.
- Lanugo is the downy hair you were covered with in the womb, from the fourth month to the time you were born.
- Vellus hair is fine, short, downy hair which grows all over your body until you reach the age of puberty.
- Terminal hair is the thicker, longer hair that grows on your head until you reach puberty — then it grows in your pubic region and under your armpits as well.
- Most men also grow terminal hair on their chins.
- The curliness of your hair depends on the shape of the follicle.

Hair colour

- Hair is red or auburn if the follicles contain the red kind of the melanin pigment. All other hair colours come from black melanin.

TEETH

Biting into food and grinding it into small and soft enough lumps for you to swallow places enormous wear on teeth. So their outer coating is made of the body's hardest substance, the white material, enamel. Inside the enamel is a softer material called dentine, but even dentine is as hard as the hardest bone. Yet even though they are tough, teeth can be eaten away by the bacteria that thrive in the mouth when you eat starchy and sugary foods. This is why it is so important to clean your teeth well, especially after eating sugary foods. Fluoride is often added to drinking water at the waterworks to protect teeth from decay. Even so, many children and adults must have holes filled by the dentist from time to time.

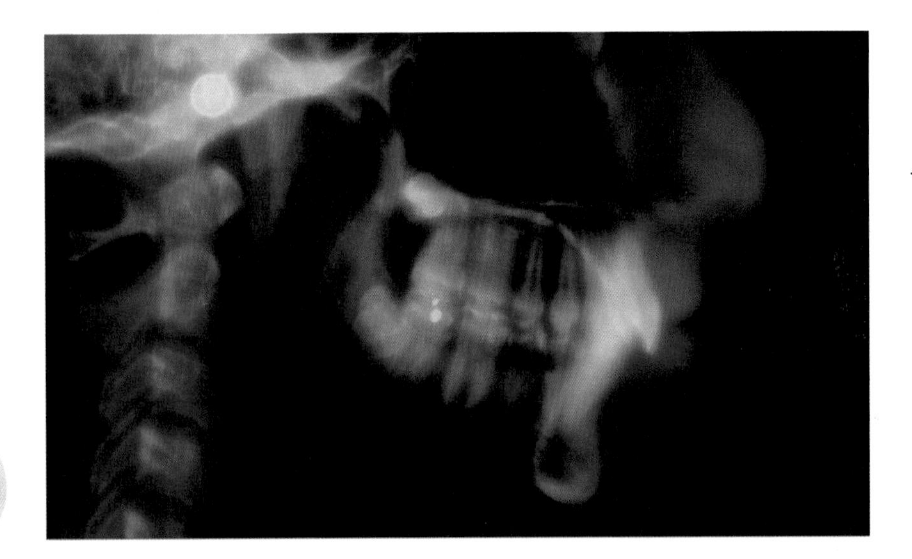

X-RAY OF ADULT JAW
Everyone has two sets of teeth during their lives. The first set of 20 are called milk teeth, and start to appear when you're about the age of six months — a process called teething. When you're about six, the milk teeth begin to be replaced by 32 adult teeth which must last you all your life. In this X-ray, the tooth roots can be seen clearly.

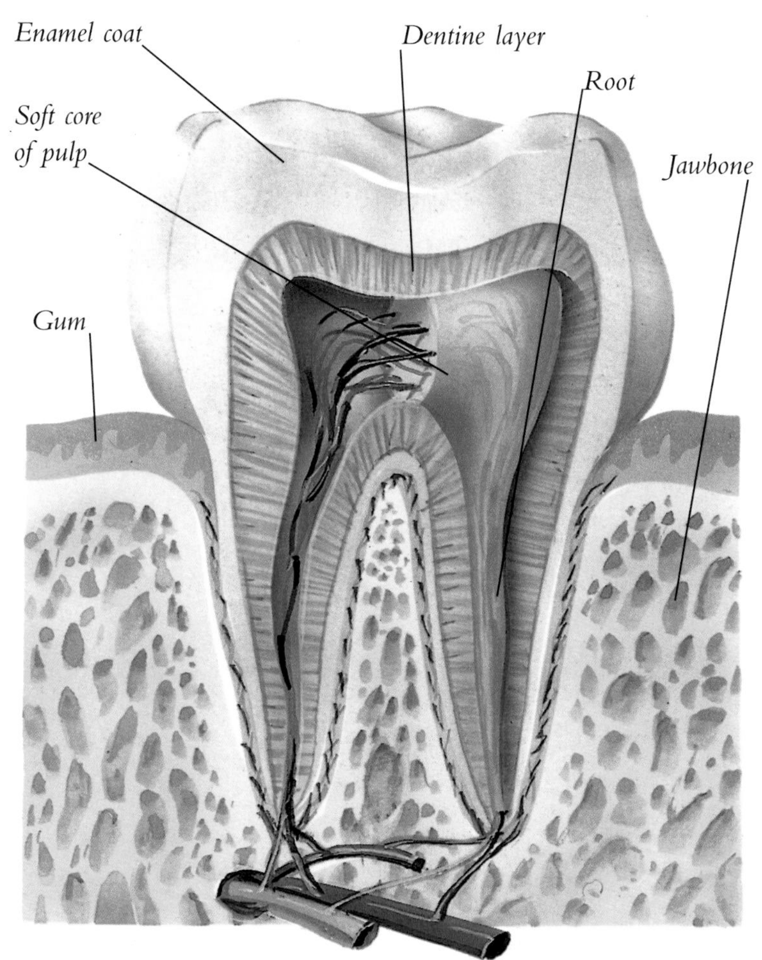

INSIDE A TOOTH

Teeth have long roots that fit into sockets in the jaw bones but they are surrounded by soft gums. At the heart of each tooth is a living pulp of blood and nerves. Around this is a layer of dentine. Around this, where it projects from the gum, is a layer of hard white enamel.

Kinds of teeth

Molars

- Molars are the big strong teeth at the back of your mouth on either side. They have flat tops with ridges and are good for grinding and softening food. You have two or three pairs of molars on each side.
- The third pair of molars, called wisdom teeth, are the last to grow, and in some people, never grow at all.

Premolars

- As an adult, you should have two pairs of premolars on each side. They have two edges but like molars they are good for grinding.

Canines

- You have just one pair of canines on each side. They are the big, pointed teeth just behind your front teeth — ideal for tearing food.

Incisors

- Incisors are the flat front teeth, with sharp edges for cutting food. There are two pairs on each side of the mouth.

THE SKELETON

Your body is held up by a strong framework of bones called the skeleton. The skeleton not only provides an anchor for the muscles, but supports the skin and other tissues, and protects your heart, brain and other organs. It is made up of over 200 rigid bones, joined together by softer, rubbery cartilage. The skeleton is the one part of the body that survives after we die, but while we are alive, it is a living tissue and is constantly being renewed as old bone cells die and new ones are born in the bone's core, or marrow.

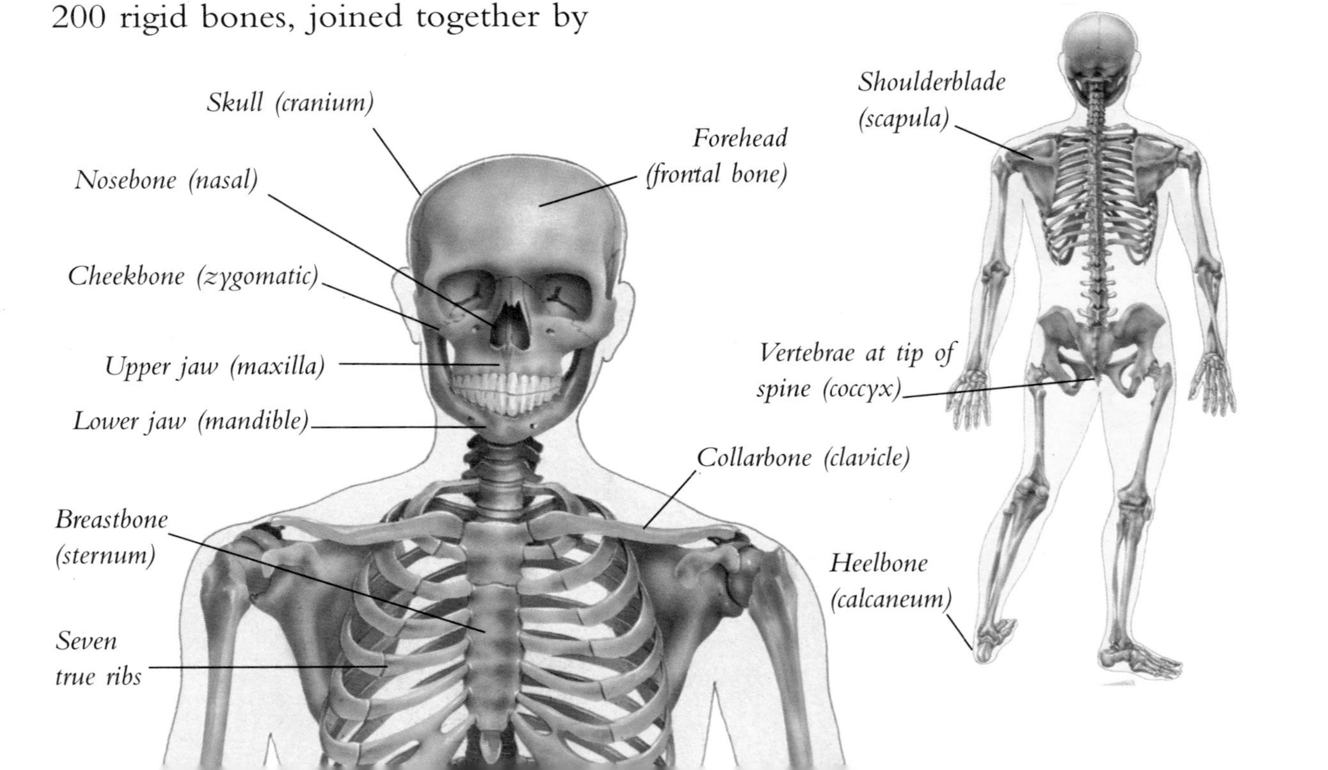

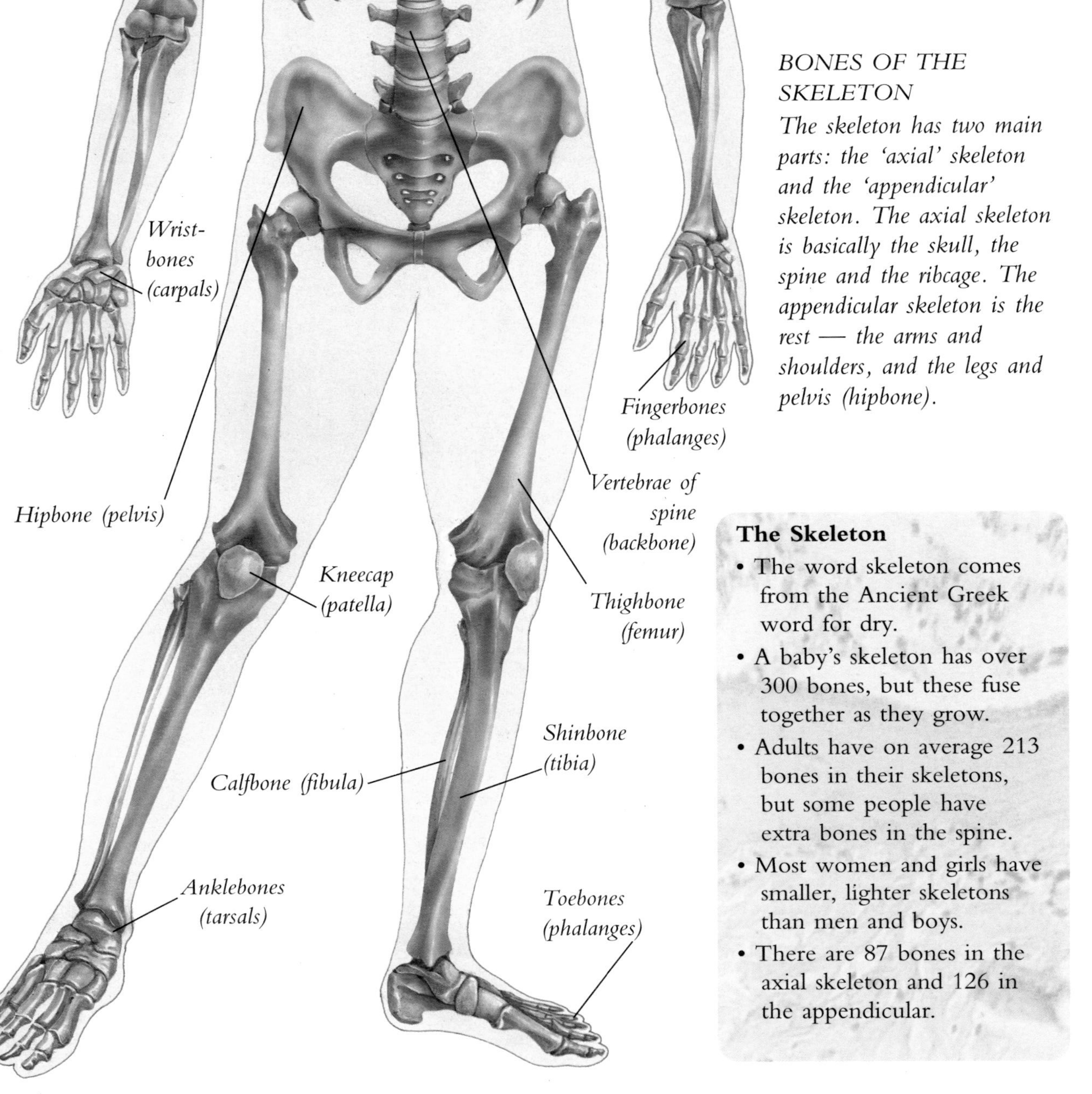

BONES OF THE SKELETON
The skeleton has two main parts: the 'axial' skeleton and the 'appendicular' skeleton. The axial skeleton is basically the skull, the spine and the ribcage. The appendicular skeleton is the rest — the arms and shoulders, and the legs and pelvis (hipbone).

The Skeleton

- The word skeleton comes from the Ancient Greek word for dry.
- A baby's skeleton has over 300 bones, but these fuse together as they grow.
- Adults have on average 213 bones in their skeletons, but some people have extra bones in the spine.
- Most women and girls have smaller, lighter skeletons than men and boys.
- There are 87 bones in the axial skeleton and 126 in the appendicular.

MUSCULATURE

Every move you make — running, dancing, smiling and everything else — depends on muscles. You even need muscles to sit still. Without them you would slump like a rag doll. Muscles are bundles of fibres that tense and relax to move different parts of the body, and there are two kinds: muscles that you can control, called voluntary muscles and muscles that you can't, called involuntary muscles. Most voluntary muscles are 'skeletal' muscles that move parts of your body when you want. Involuntary muscles are like those of the heart which work automatically.

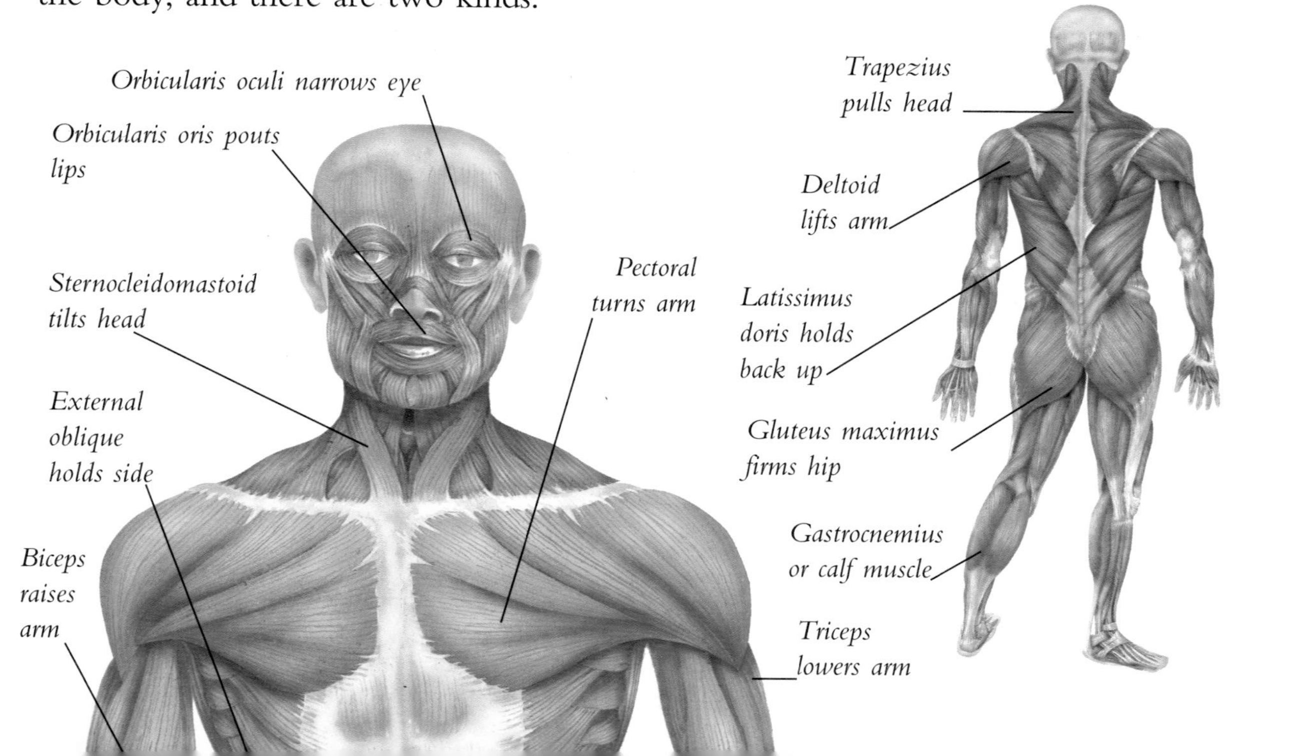

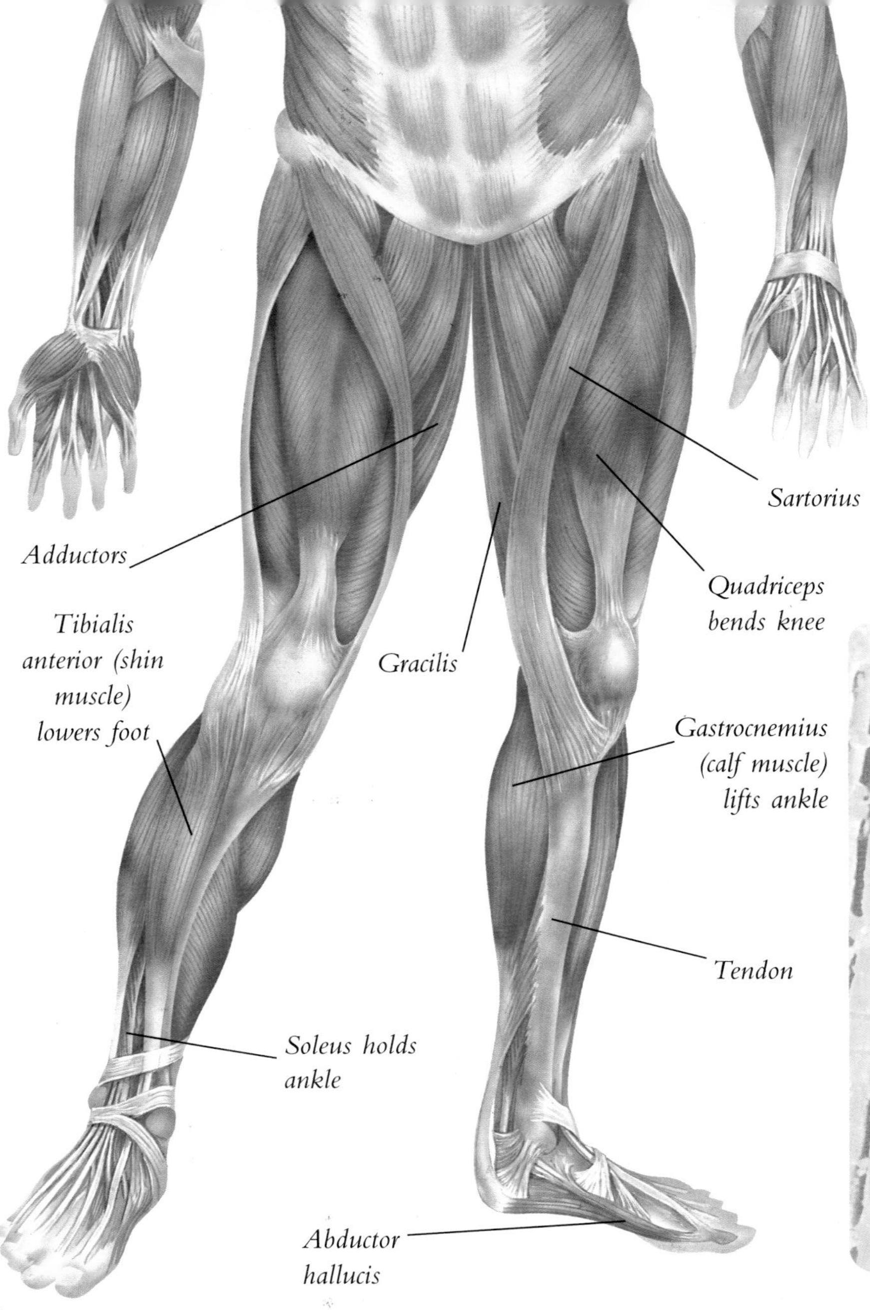

BODY MUSCLES
The body is covered with skeletal muscles, making up 40 per cent of the body's weight. There are more than 640 all told, but only those on the surface are shown here. There is another layer, or even two, beneath. In theory, you can control each muscle pair individually. But most work in combinations so well established by habit that they always work together.

Musculature

- The body's longest muscle is the sartorius in the inner thigh. The widest is the external oblique on the side of the upper body.
- The body's biggest muscle is the gluteus maximus in the buttock which can weigh 1kg or more.
- Most muscle fibres are 3cm long on average.
- Some muscles are long and bulge in the middle. Some are triangular, like the trapezius in the upper back. Some are sheet-like, like the external oblique.

DIGESTIVE SYSTEM

The body needs food in the form of small, simple molecules that can be delivered to cells in the blood. Yet the food you eat comes in big lumps and large complex molecules. So the body has its own food refinery for breaking food down into the right kind of molecules. This food breakdown refinery is known as the digestive system. It is essentially a long tube through the body, called the alimentary canal, through which the food slowly passes and is gradually digested — that is, broken down into small molecules and absorbed into the bloodstream.

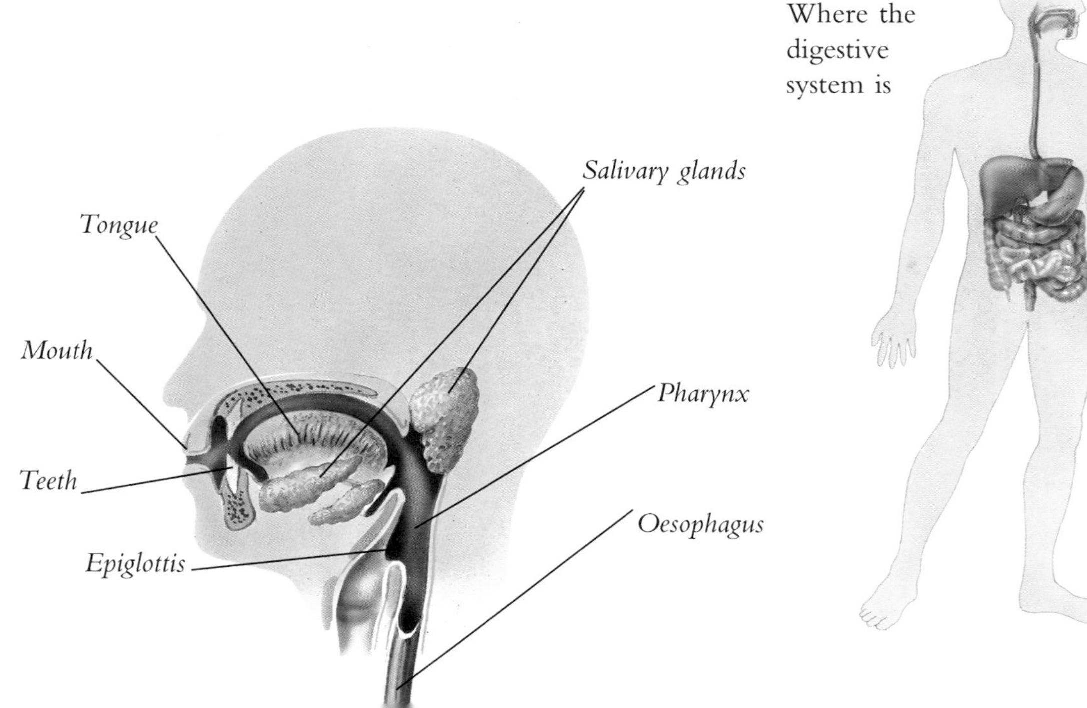

Where the digestive system is

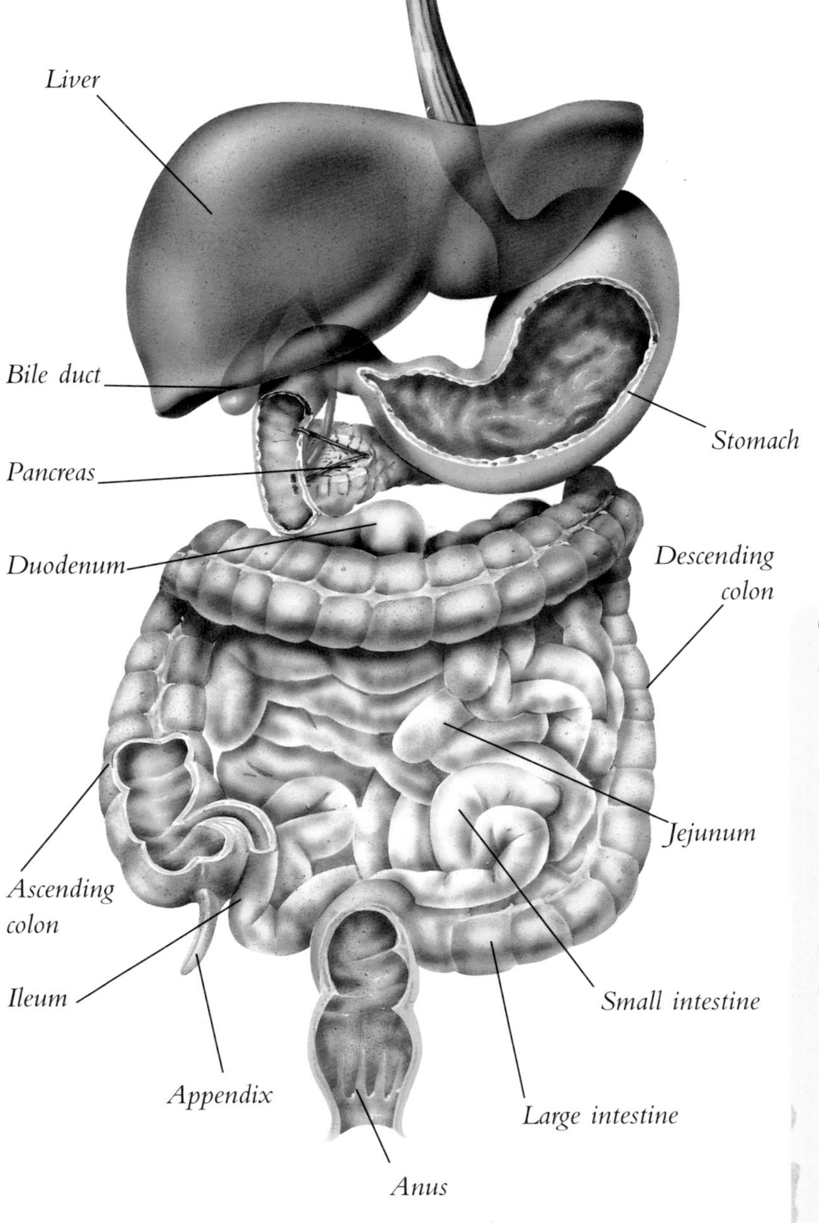

THE DIGESTIVE SYSTEM
The digestive system is essentially the alimentary canal, plus the liver and pancreas. The alimentary canal begins in the mouth, runs down through the oesophagus, or throat, then the stomach, the small intestine and the large intestine and finally ends at the anus.

The digestive system

- The alimentary canal is more commonly known as the gut.
- Your gut is folded over many times. If you could lay it out straight, it would be nearly six times as long as you are tall.
- When you are sick, your diaphragm and the muscles of your abdomen contract to squeeze partly digested food from your stomach out through your mouth. Vomit tastes bitter because of the acidic stomach juices it contains.

HOW THE LIVER WORKS

The liver is the body's biggest organ — and one of its cleverest, performing miracles of chemical processing on many different substances. It helps purify the blood, for instance, sweeping out all the tired old red cells and poisons such as alcohol. It does not actually make blood cells, but it makes vital proteins for blood plasma and is also a source of bile, the fluid that helps dissolve fat in food.

The liver's most important task, though, is to receive chemicals digested from food, and repackage them for use all around the body when needed. It takes carbohydrates, for instance, and turns them into glucose, the number one fuel of body cells. Guided by two chemical messengers, glucagon and insulin, the liver helps ensure our blood always has the right level of glucose — without which we would fall into a coma and quickly die.

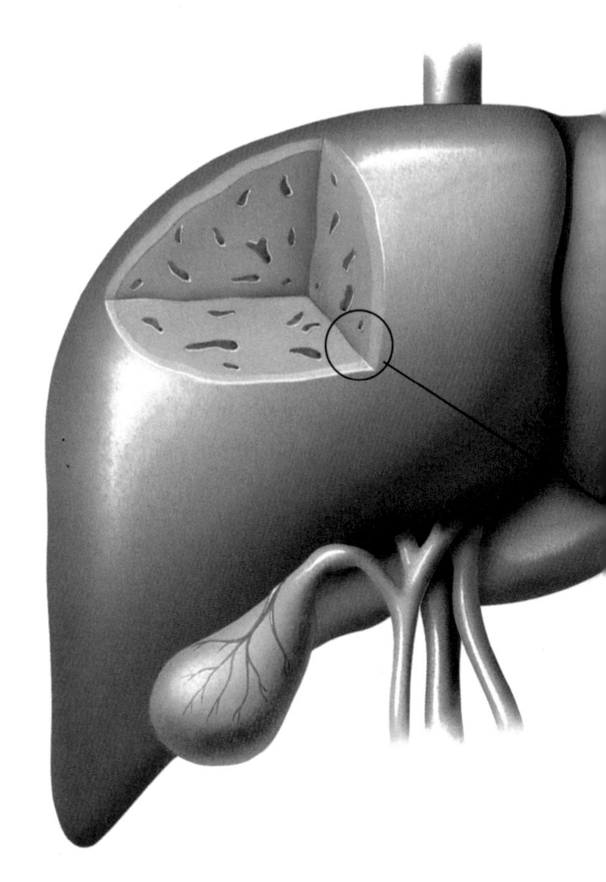

LIVER
The liver is a large organ sitting just above the stomach. Despite the huge range of tasks it performs, it is actually quite simply made. Its jelly-like interior is made up from thousands and thousands of hexagonal lobes, or 'lobules'.

LIVER LOBE
The key to the liver's operations are the box-shaped 'lobules', with sections like the segments of oranges. Blood from the liver's twin supplies flow into each segment from the outside edge, through a channel called a sinusoid, and out through a vein in the centre. There are special liver cells or 'hepatocytes' lining the sinusoid, and as the blood flows through it, these cells extract the right chemicals, process them and return them to the blood — except for bile, which is sent out the back.

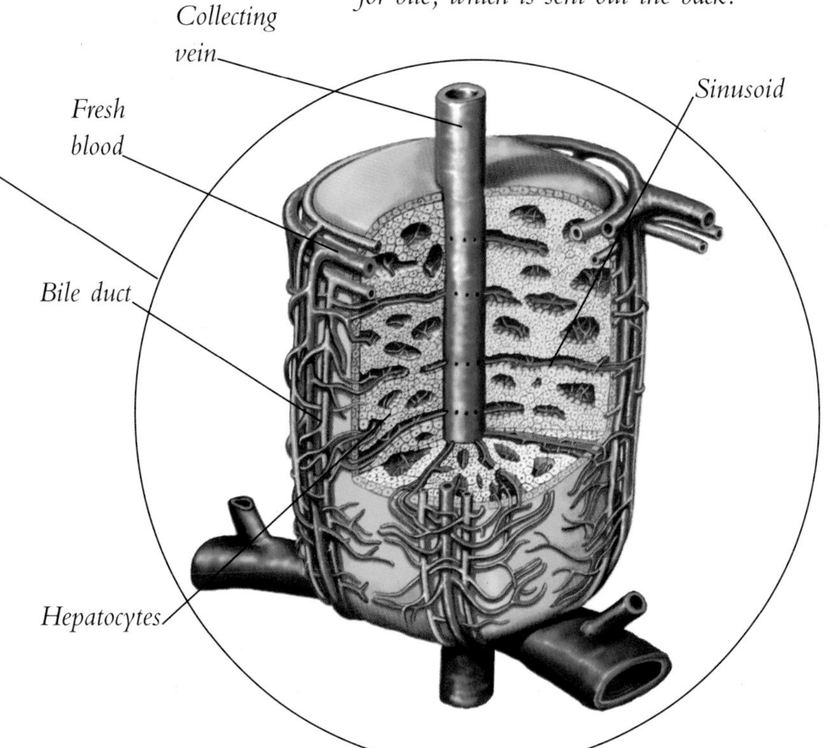

Liver facts

- The word 'hepatic' means belonging to the liver.
- Uniquely, the liver has two blood supplies. One is an artery, the hepatic artery; the other is a vein, the portal vein. Both enter the liver through its 'portal', or gate.
- The third pipe passing through the liver's portal is the common 'bile' duct, which carries bile, the acid the stomach uses to help break fatty food down.
- The liver acts as a despatch centre for the chemicals extracted from food by the digestive system.
- The liver is the body's prime energy store, holding glucose in the form of the chemical glycogen.
- The liver packs off excess food energy for long term storage as fat.
- The liver breaks down proteins and stores vitamins and minerals.
- The liver clears the blood of old blood cells and makes new plasma and proteins.

INSIDE THE KIDNEYS

Water is crucial to the working of the body, and the kidney is the key to water control. It holds water back when needed, or lets it run out as urine when there is too much. The water is mixed in with the blood, so the kidney has to draw off water without losing any of the blood's vital ingredients. At the same time, it must clear the blood of poisonous waste. The kidneys are basically a highly efficient filtration unit, cleaning the blood as it washes through — catching larger materials and letting smaller blood ingredients pass through to the next stage. The kidney then reabsorbs the wanted ingredients and water, and allows toxic waste and unwanted water to flow away in the urine.

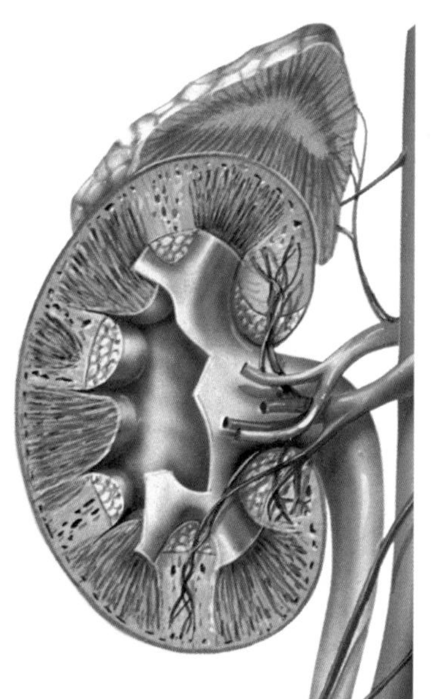

KIDNEY
The kidneys are a pair of bean shaped organs in the small of the back, along the body's main arteries and veins. Blood entering the kidney through an artery is distributed through a million or so filtration units called nephrons.

NEPHRON
A nephron is an intricate network of little tubes called convoluted tubules, wrapped round by an even more intricate network of blood capillaries. It is in these capillaries that wanted ingredients are reabsorbed into the blood.

Glomerulus

Blood in and out of nephron

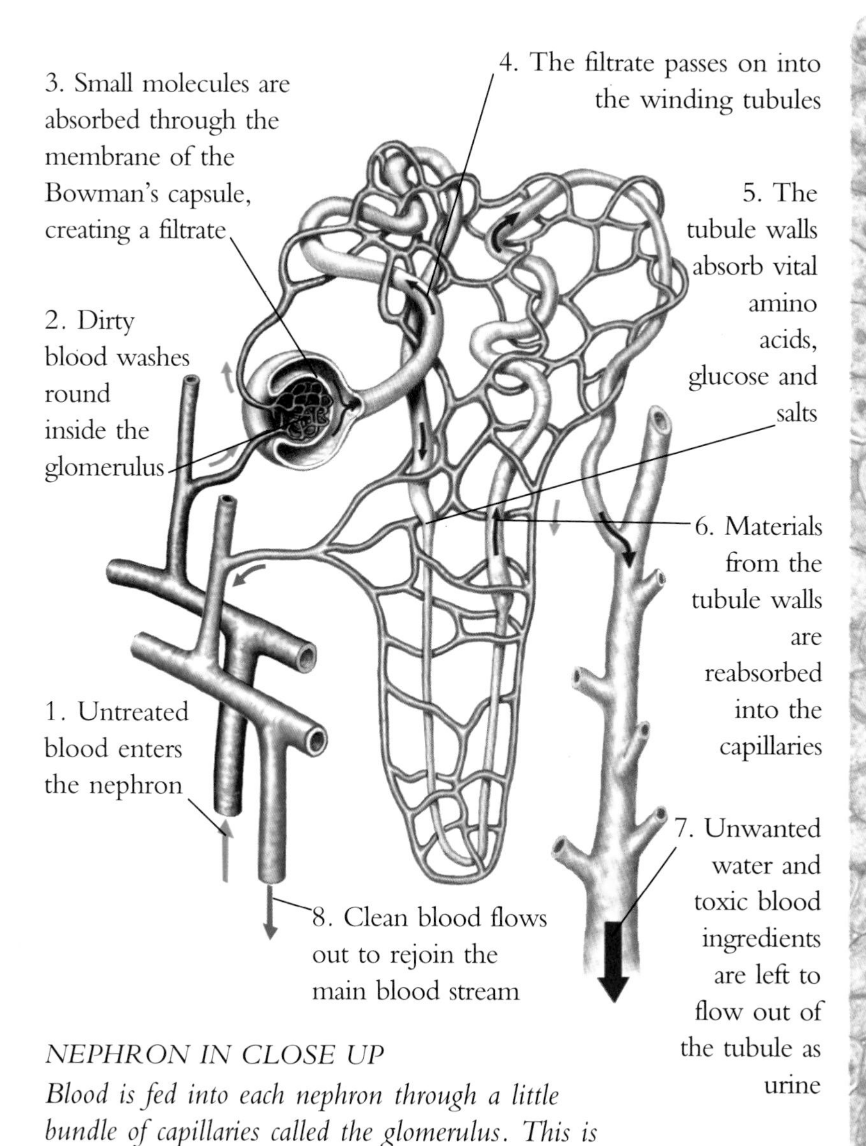

NEPHRON IN CLOSE UP

Blood is fed into each nephron through a little bundle of capillaries called the glomerulus. This is held in a cup called the Bowman's capsule. Between the two there is a thin membrane, and this membrane is the kidney's filter.

Kidneys

- Blood flows through the kidneys at 1.3 litres a minute.
- All the body's blood flows through the kidneys in just 10 minutes, so the blood is filtered by the kidneys hundreds of times a day.
- The kidneys absorb 170 litres of filtrate from 2,000 litres of blood everyday.
- The kidneys only let go 1.2 litres of urine from every 2,000 litres of blood, reclaiming all the rest.
- Only small molecules get through the membrane of the Bowman's capsule — salt, water minerals, glucose, urea and creatine.
- Urea is the waste from the breakdown of proteins in the liver; creatine is the waste from muscle action.
- The tubule walls retrieve all of the vital amino acids and glucose from the filtrate for returning to the blood, and 70 per cent of the salt.

THE WATER SYSTEM

Your body needs one more vital input as well as oxygen and food — water. You can live for a month without food but only a few days without water. If the body's water content goes up or down by 5 per cent, the results are disastrous. The urinary system plays a key role in keeping the amount of water in the body steady.

You gain water by drinking, eating and as a by-product of cell activity. You lose it by sweating, breathing and through urinating. The amount you gain and lose through cell activity, sweating and breathing stays much the same. So your body controls water by balancing the amount you drink against the water you lose in urine.

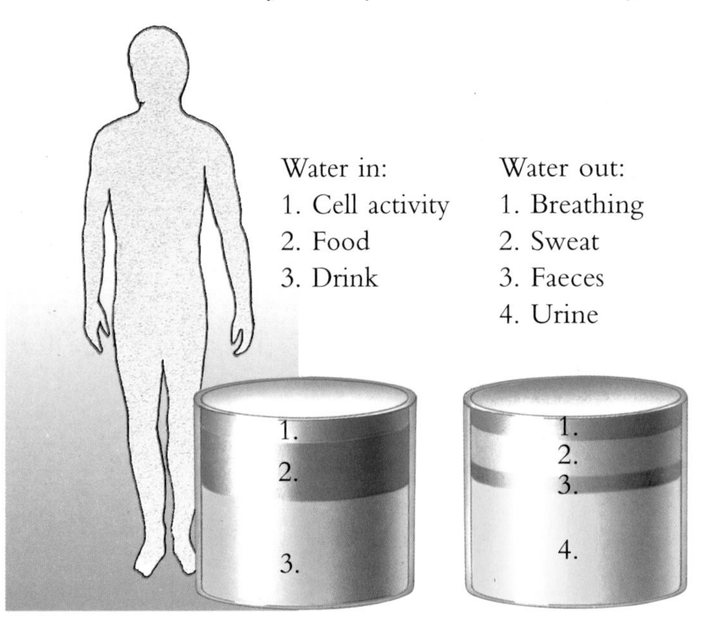

WATER IN THE BODY
The human body is over 60% water — and it is crucial that the balance of water in the body remains steady. Typically, you will take in 2.2 litres of water a day, 1.4 in drinks and 0.8 in food. Body cells add an extra 0.3, bringing the total to 2.5 litres. So, to keep you from being swamped, your body must also lose 2.5 litres. Typically, 0.3 goes out as vapour on your breath, 0.5 in sweat, 0.2 in your faeces and 1.5 in urine.

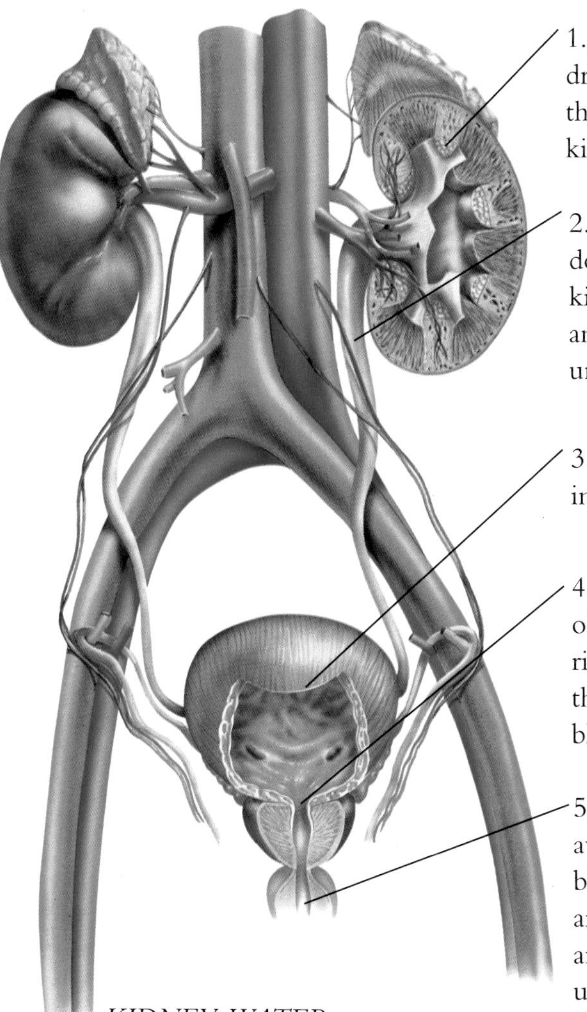

KIDNEY WATER

The urinary system drains unwanted water drawn off from the blood by the kidneys. It pipes the water or 'urine' down to the bladder where it builds up until the pressure of water there makes you want to urinate.

Water and salts

How the kidneys control water

- The amount of water the kidneys let out as urine depends on the amount of salt dissolved in the blood.
- If you drink a lot, the water in the blood becomes very diluted, and the kidneys let lots of water out, giving pale and watery urine.
- If you drink only a little or sweat a lot, the kidneys hold on to more water, and the urine is stronger.

Why water balance matters

- There is slightly more potassium salt dissolved in the water inside body cells than in the water outside.
- There is slightly more sodium chloride (common salt) dissolved in the water outside cells.
- Water seeps through cell walls from areas of low salt to areas of high salt in a process called osmosis.
- If the body water gets too dilute, it seeps into cells and swells them up. If it gets too salty, cell water leaks out.

Nervous System

Like a busy telephone network, your nervous system is continually buzzing with activity, whizzing messages to and fro all over the body. All the time, millions of nerve impulses reach the brain from the body's sense receptors, and just as many leave the brain with instructions for muscles to move or organs to work.

The heart of your nervous system is your brain and the bundles of nerves running down your spine, known as the spinal cord. Together, the brain and spinal cord make the central nervous system or CNS. Nerves spread out from the CNS to all parts of the body. These nerves are the Peripheral Nervous System or PNS.

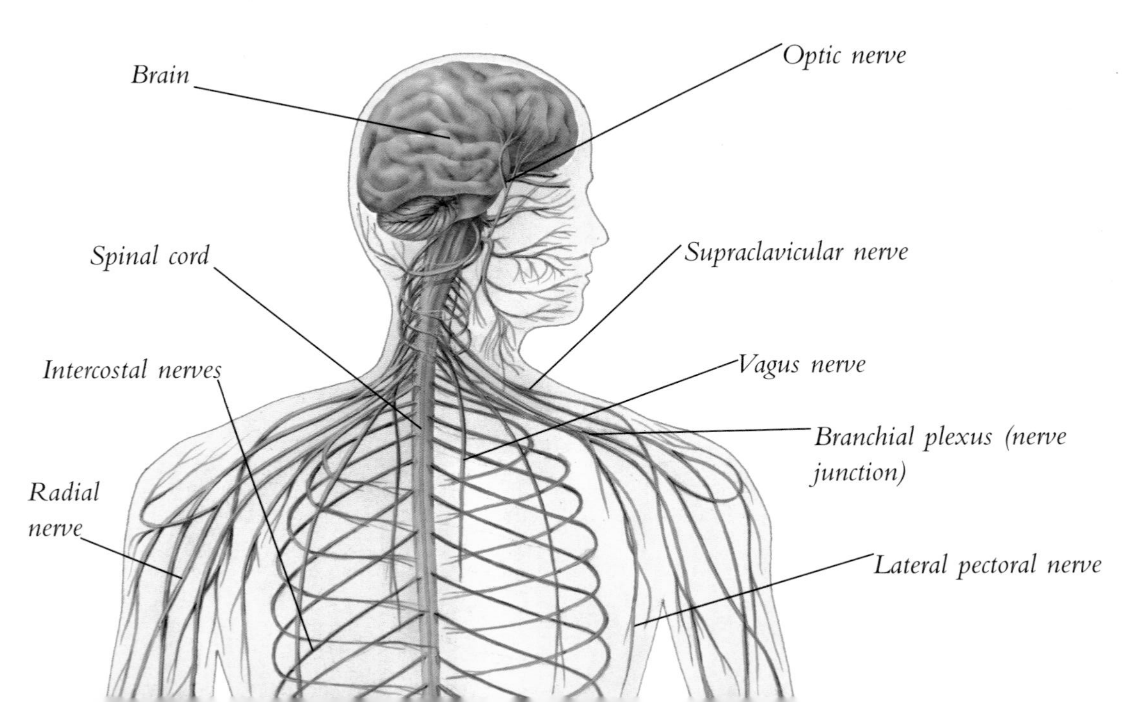

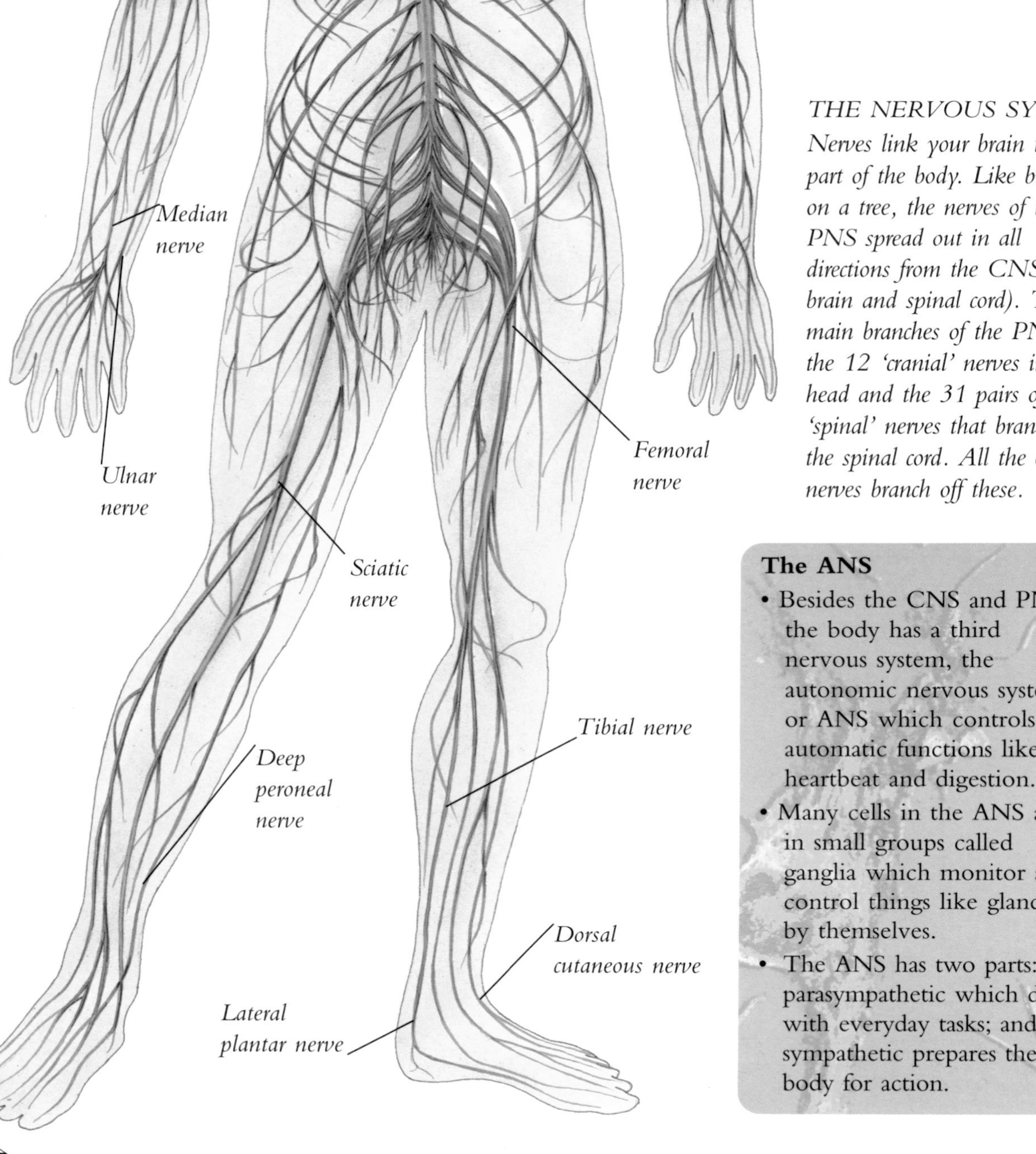

THE NERVOUS SYSTEM
Nerves link your brain to every part of the body. Like branches on a tree, the nerves of the PNS spread out in all directions from the CNS (the brain and spinal cord). The main branches of the PNS are the 12 'cranial' nerves in the head and the 31 pairs of 'spinal' nerves that branch off the spinal cord. All the other nerves branch off these.

The ANS

- Besides the CNS and PNS, the body has a third nervous system, the autonomic nervous system or ANS which controls automatic functions like heartbeat and digestion.
- Many cells in the ANS are in small groups called ganglia which monitor and control things like glands by themselves.
- The ANS has two parts: the parasympathetic which deals with everyday tasks; and the sympathetic prepares the body for action.

THE BRAIN

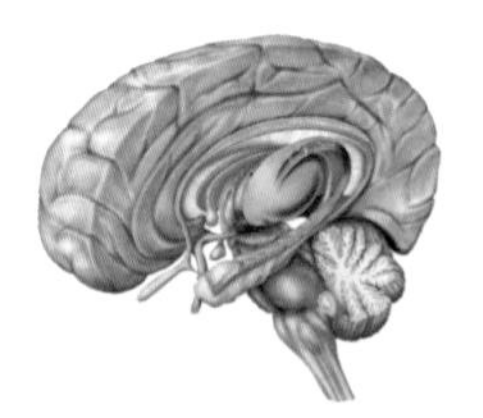

Inside your head you carry the most amazing structure in the universe: a human brain. It looks like little more than a large, soggy grey walnut with its wrinkled surface. But within this soft mass are billions of inter-linked nerve cells. The chemical and electrical impulses whizzing through all these cells create all your thoughts, record every sensation and control nearly all your actions. Every second of your life, your brain is receiving signals from the rest of your body and issuing instructions via the body's network of nerves. Even more amazingly, it lets you go on thinking, even when there are no inputs or outputs.

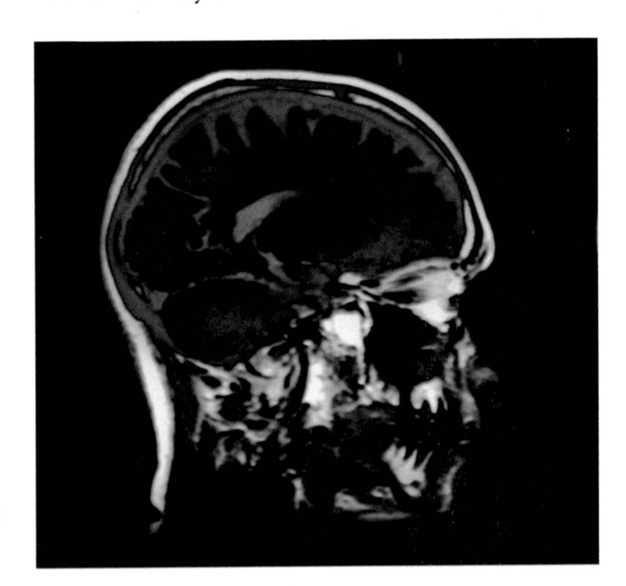

SEEING THE BRAIN
This model of a head with the top of the skull cut off (right) shows the brain's two halves or 'hemispheres' and the wrinkles that cover its surface. But scientists can only see inside a working brain with the aid of special scanners which reveal brain activity (left) with traces of chemicals injected into the blood.

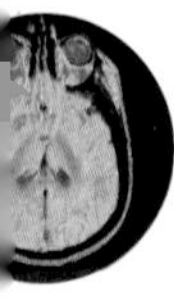

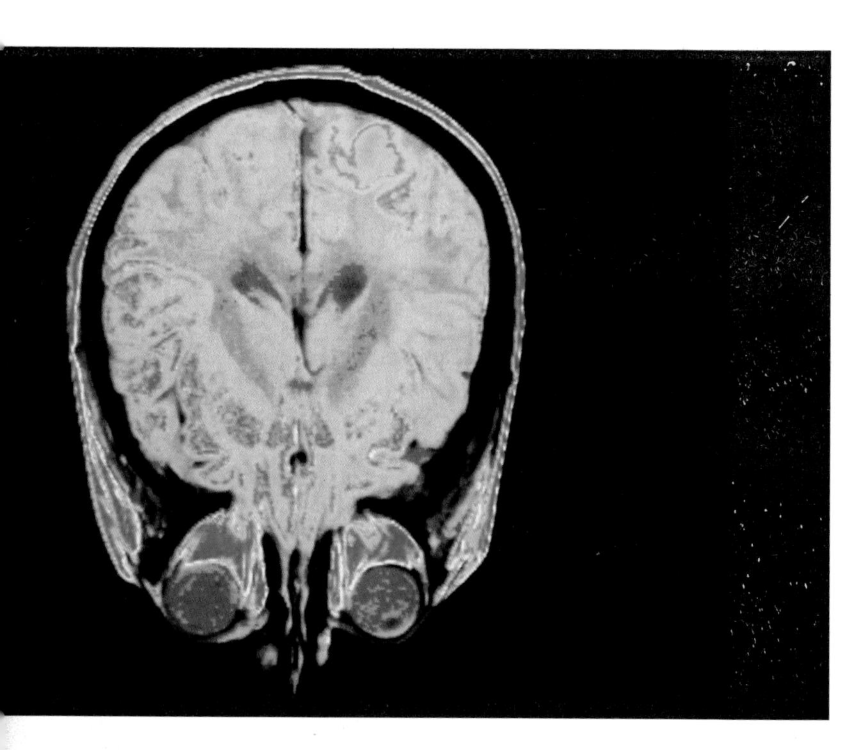

BRAIN SLICE
This scan shows a 'slice' through a living brain from above in computer colours. The yellow mass is the brain, the blue the skull and the red discs are the eyeballs.

VULNERABLE BRAIN

Even though it is no bigger than a bag of sugar, the brain's buzzing activity demands huge amounts of energy — and brain cells also depend on oxygen in the blood. If the blood supply to your brain were cut off, you would lose consciousness within ten seconds, and die within minutes.

Brain facts

- There are 100 billion cells in your brain, each connected to as many as 25,000 others — so there are 25,000 times 100 billion nerve connections in your brain, or two and a half million billion.
- Girls' brains weigh, on average, 2.5 per cent of their body weight (about 1.25kg); boys' brains weigh 2 per cent of their body weight (1.4kg).
- The brain makes up under 2% of your body's weight, yet demands over 25% of it's blood supply.
- The membrane between brain cells and the blood supply is called the blood-brain barrier.
- The blood-brain barrier lets through only the finest particles, including glucose.
- The layout of the human brain and nervous system is much the same as other animals — but the top of the brain, the cerebrum, is much, much bigger.

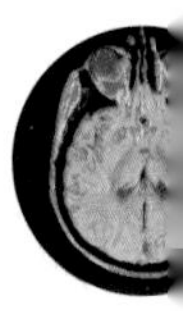

THE EYES

The picture formed on the back of your eye is a few millimetres across, yet it seems so big that it hardly occurs to you that it is just a picture, like a camera picture.

Your eyes combine the optical quality of the best cameras with astonishing versatility. A good camera may match the eyes for sharpness, but none can focus on both a speck of dust and distant galaxies. What's more, your eye works in both starlight and sunlight, a difference in brightness of a hundred billion times. It does this by adjusting its sensitivity to suit the light — which is what happens when your eyes gradually become accustomed to a dark room.

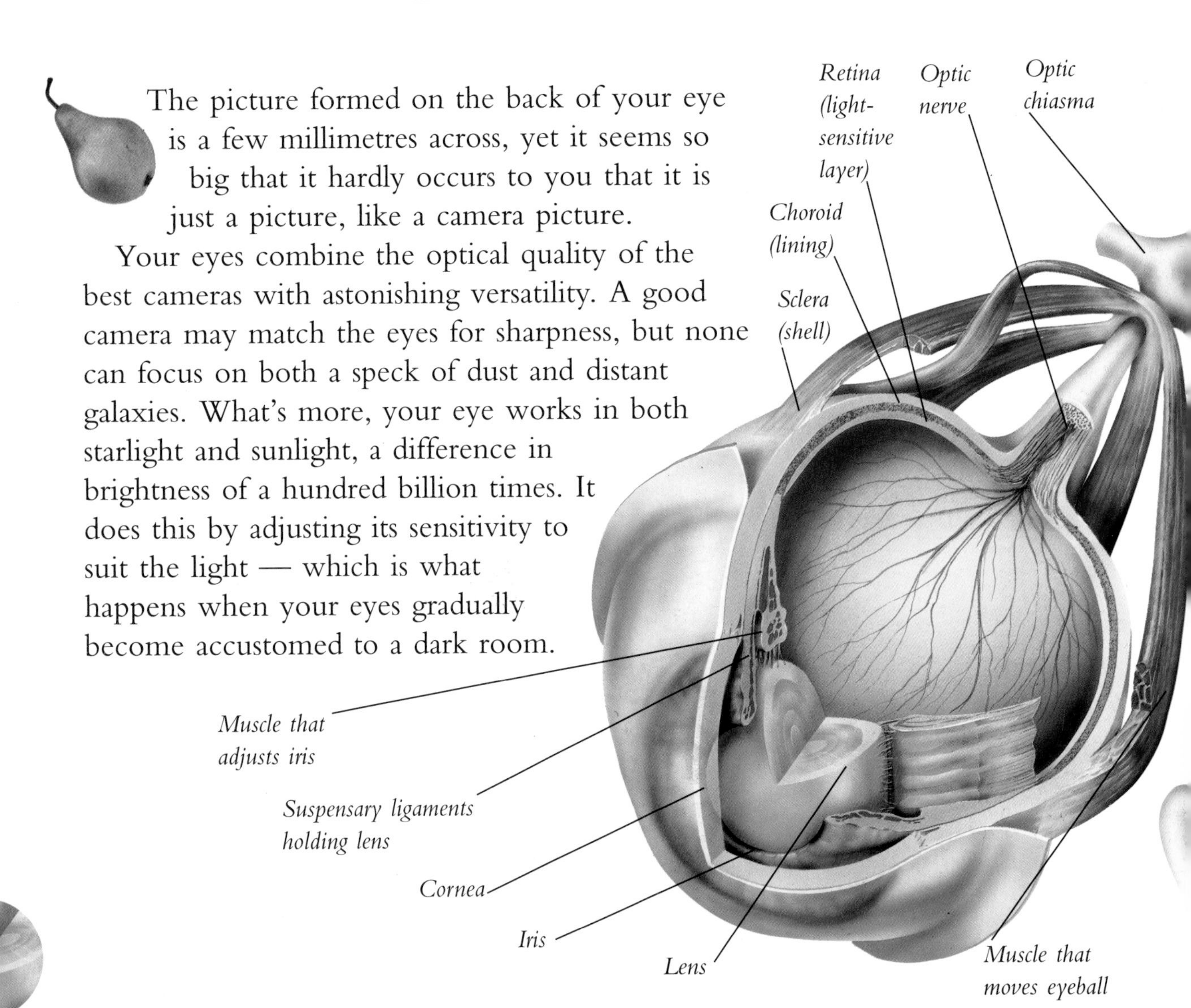

RODS AND CONES

The retina at the back of the eye is made of millions of light-sensitive cells called rods and cones, which send signals to the brain along the optic nerve ('optic' just means for eyes). The process is described in the panel.

Ganglion cell

Bipolar cell

Retina

Rods and cones

THE EYES

Your eyes are two tough little balls filled with a jelly-like substance called vitreous humour. Each is a bit like a video camera with a lens called a cornea at the front — the dark circle of your eye. This projects a picture on the back of your eye, called the retina. There is another lens behind the cornea, but this simply adjusts the focus of the cornea's picture.

Tear duct

Pupil

From eye to brain

- Whenever hit by enough light, each of the rods and cones in a little group sends a tiny electrical signal to their group leader, the bipolar cell.
- Instantly, the bipolar cell passes the message on to a ganglion cell.
- The ganglion is always firing signals down to the optic nerve. But the signals from the bipolar excite it to fire faster.
- Signals from all the ganglion cells in each eye whizz away down the dual carriageway of your two optic nerves. They meet at a crossroads in the middle of your brain called the optic chiasma.
- At the optic chiasma, one half of each carriageway goes off to the right of the brain and the other to the left, so each half of the brain gets half the picture from each eye.

The Ear

Sounds are just tiny vibrations in the air, and your ears are incredibly sensitive devices for picking up these tiny vibrations.

The flap of skin on the side of your head is only the entrance to the real ear, and simply funnels the vibrating air towards sensitive pressure detectors inside the head. The ear actually has three sections. The outer ear is the earflap and the ear canal, the tunnel into your head. Inside your head, in the 'middle ear', the sound hits a taut wall of skin called the eardrum, shaking it rapidly.

As it shakes, it rattles three tiny bones, or ossicles. Even further inside, in the 'inner ear', is a curly tube full of fluid, called the cochlea. As the ossicles vibrate, they knock against this tube, making waves in the fluid. Minute hairs waggle in the waves, sending signals along nerves to the brain.

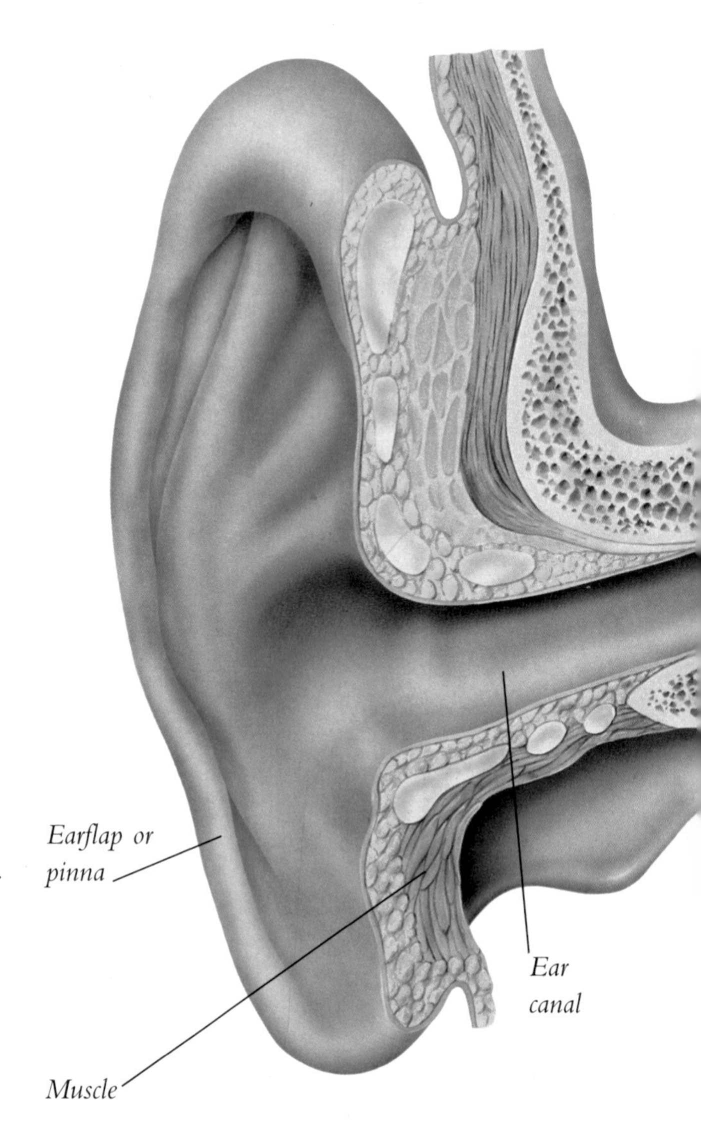

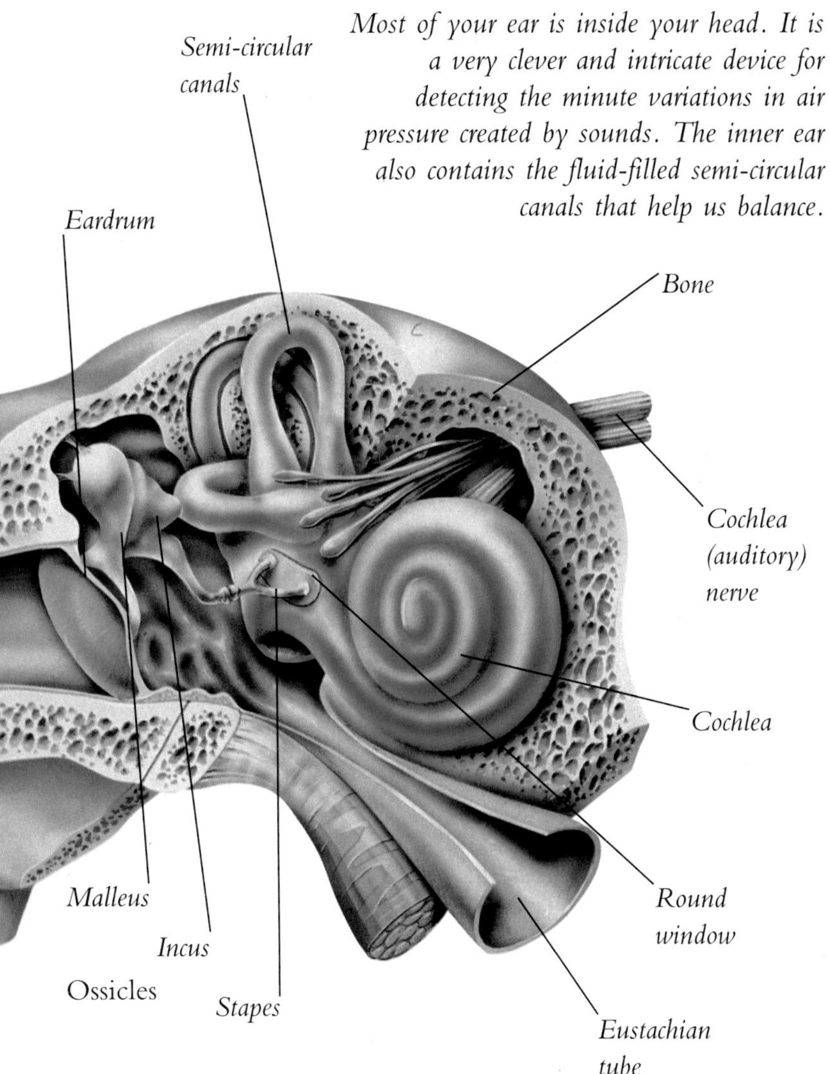

THE EAR
Most of your ear is inside your head. It is a very clever and intricate device for detecting the minute variations in air pressure created by sounds. The inner ear also contains the fluid-filled semi-circular canals that help us balance.

Hearing

- Because you have two ears, you can pinpoint where a sound came from, though not always accurately.
- You can pinpoint sound because a sound to the left of you is slightly louder in the left ear than in the right ear and vice versa.
- If your hearing is normal, you can hear sounds as deep as 20 Hz, deeper than a bass drum.
- If your hearing is normal, you can hear sounds as high as 20,000 Hz.
- Sound intensity is measured in decibels on a geometric scale — that is, three decibels are twice as loud as two, and four twice as loud as three.
- If your hearing is normal, you can hear sounds as quiet as 10db (decibels), which is quieter than leaves rustling in trees.
- If your hearing is normal, you can hear sounds as loud as 140db or more, which is as loud as a jet engine close up.

TOUCH

Touch is the most widely spread of all your senses, with receptors all over your body, from head to toe. Some places have many receptors, such as your hands and face. Others, like the small of the back, have comparatively few.

The touch receptors in the skin react to four main kinds of sensation: a light touch, continuous pressure, heat and cold, and pain. All four of these sensations are felt in skin areas where there are only free nerve endings as receptors. But in some places there are a handful of different kinds of specialized receptors too, and each seems to respond to one kind of sensation more strongly.

There are five main specialized receptors, each named after its discoverer. Two respond mostly to a sudden touch or knock: Pacini's corpuscles and Meissner's endings. The other three respond mostly to steady pressure: Krause's bulbs, Merkel's discs and Ruffini's endings.

When a receptor in the skin is stimulated, it fires nerve signals to the brain. The rate the nerve fires signals tell the brain, for instance, how heavy the touch is or how cold it is.

FINGER READING

In many blind people, sensitivity to touch is so highly developed that they can read with their fingers. The Braille system gives letters as different patterns of raised dots. Blind people can even operate computers using braille.

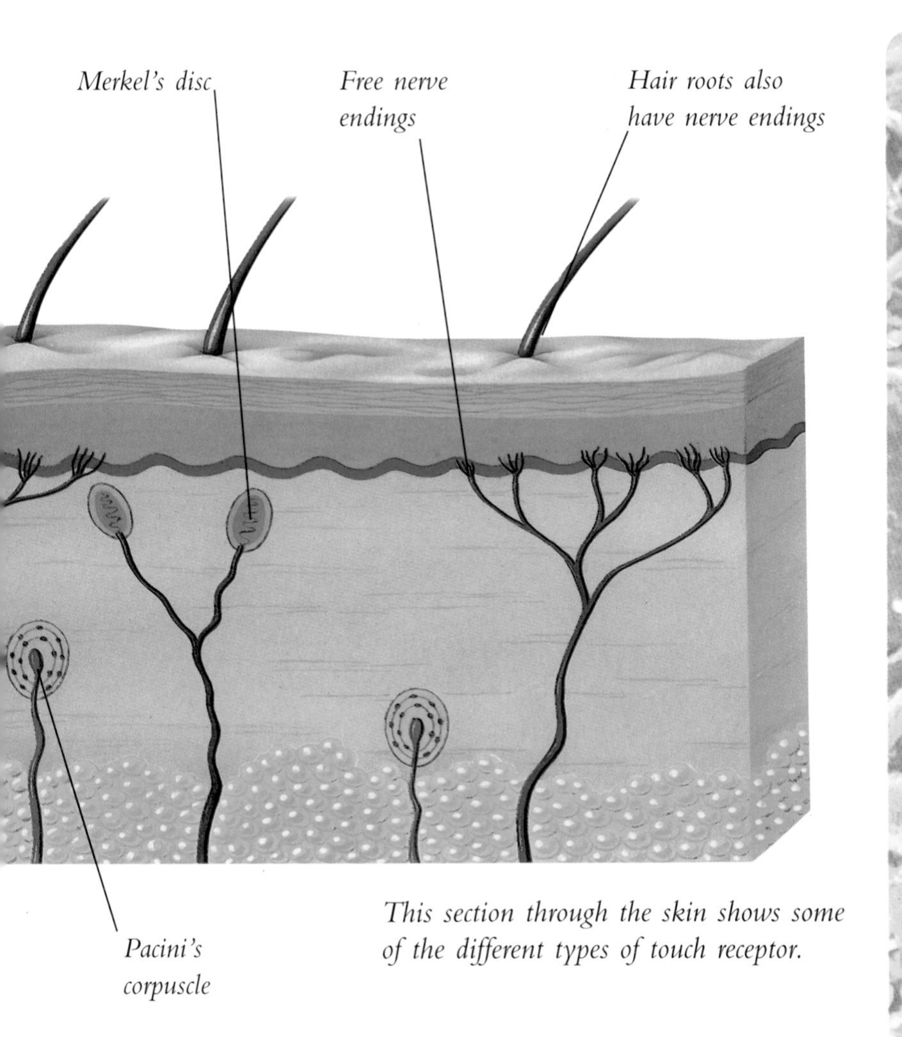

This section through the skin shows some of the different types of touch receptor.

But the receptor does not go on firing forever. Instead, the firing rate falls off as it adapts to the stimulus. Once it has alerted the brain, it need only send an occasional reminder — which is why you cease to feel clothes soon after you put them on in the morning.

Special touch receptors

Sudden pressure

- Onion-shaped Pacini's corpuscles are the biggest and deepest of the touch receptors. Pacini's corpuscles respond to very heavy pressure and fast vibrations.
- Meissner's endings are little capsules just below the surface of the skin, especially on the palm side of the fingers. They detect light touch and vibrations and may be what enable you to distinguish delicate textures by feel alone.

Steady pressure

- Merkel's discs are bowl-shaped receptors that respond to continuous pressure.
- Krause's bulbs are bulb-shaped receptors deeper in the skin that are believed to respond to cold as well as steady pressure.
- Ruffini's endings probably respond to changing temperature as well as continuous pressure.

SMELL

The human nose is not quite so sensitive as that of some animals. But even so, your nose detects smells just by picking up a few airborne molecules of a substance, and it can distinguish between over 3,000 different chemicals.

Our sense of smell relies on a small patch inside the top of the nose called the olfactory epithelium. This patch is just 6cm square, yet it is packed with 10 million olfactory (smell) receptors. These receptors feed into a small area called the olfactory bulb, which then relays messages to the brain.

SENSE CENTRE

The nose is divided into two halves by a wall of cartilage called the nasal septum. But both nostrils lead smells up to the olfactory patch high inside the nasal cavity. Here they stimulate tiny hair-like cells called cilia on the top of millions of olfactory receptor cells.

THE SCENT OF TULIPS

When you smell the sweet scent of flowers, your nose is actually reacting to small molecules given off by the flowers and wafted in the air up your nostrils.

PICKING UP A SCENT

On the top of each olfactory receptor are 20 or so tiny hair-like 'cilia', and it is these which are stimulated by molecules of a smelly substance. For a smell to be detected, it must vaporize so that it can be wafted up the nostrils. It must then dissolve in water so it can get through the thick coat of mucus covering the cilia. The number of receptors stimulated tells you just how strong a smell is. The position of the cilia stimulated may be what helps your brain identify the smell.

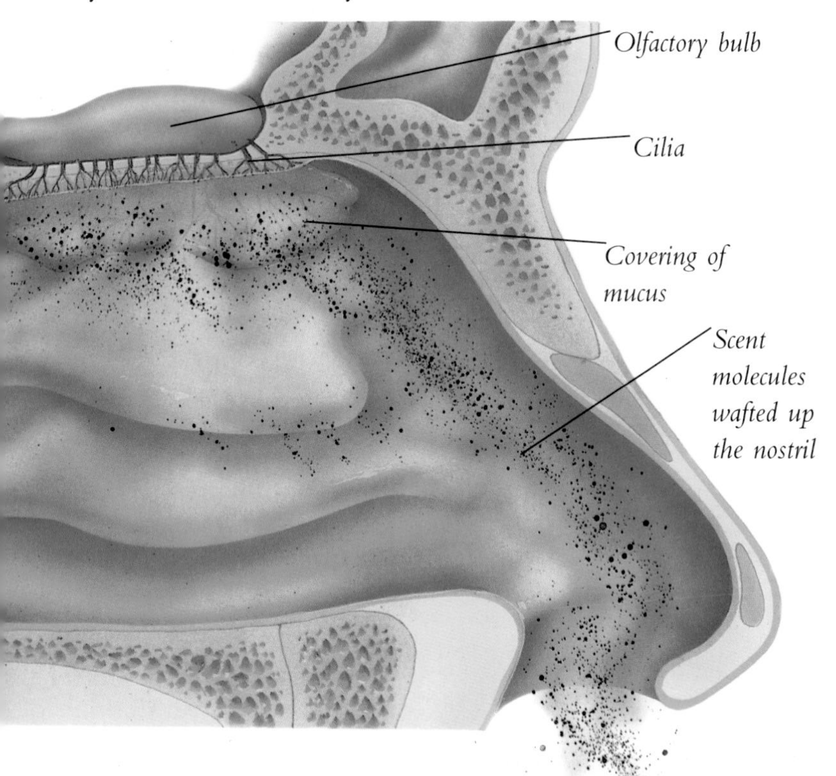

Smell

- Nasal receptors react more to change than to steady stimuli. The nasal receptors may soon stop responding to a smell so that you have to sniff hard to go on smelling a faint scent.
- Most of the chemicals you can smell contain at least three atoms of carbon.
- The sense of smell is strongest in babies, and helps a baby recognize its mother.
- By the age of 20, your sense of smell will have dropped off by 20%.
- By the age of 60, your sense of smell will have dropped off by 60%.
- The smell of chemical 'pheromones' in sweat may play a part in the attraction between men and women.
- Dogs have olfactory patches 30 times as big as humans.
- Humans can detect smells in concentrations of 1 part in a billion. Dogs are 10,000 times more sensitive.

TASTE

Our sense of taste is in some ways much vaguer than all the other senses, and we can only taste the difference between sweet food, salty food, sour food and bitter food. However, when we eat we bring a whole range of other sensations into play — heat and cold, texture, look and especially smell. So most of us can tell the difference between a huge range of foods, and trained wine and food tasters can detect and appreciate small variations in taste.

The body's taste receptors are the taste buds of the tongue, set in tiny pits. To reach them, food must first be dissolved in saliva, which is why you can't at first taste food which doesn't dissolve quickly. There are four kinds of bud, each responding to a different taste: sweet, salty, sour or bitter. When any is set off, it sends a message to the brain.

CHEESY FLAVOURS
Our taste buds can detect very little real difference between all these cheeses. But other senses such as smell combine with taste to reveal the rich differences in flavour which makes certain cheeses so popular.

TASTE BUD
In each taste bud there is a cluster of cells with tiny hairs on the end. These hairs are washed over by saliva pouring into the bud. If the right taste is in the saliva, the hair triggers off the receptor cell below.

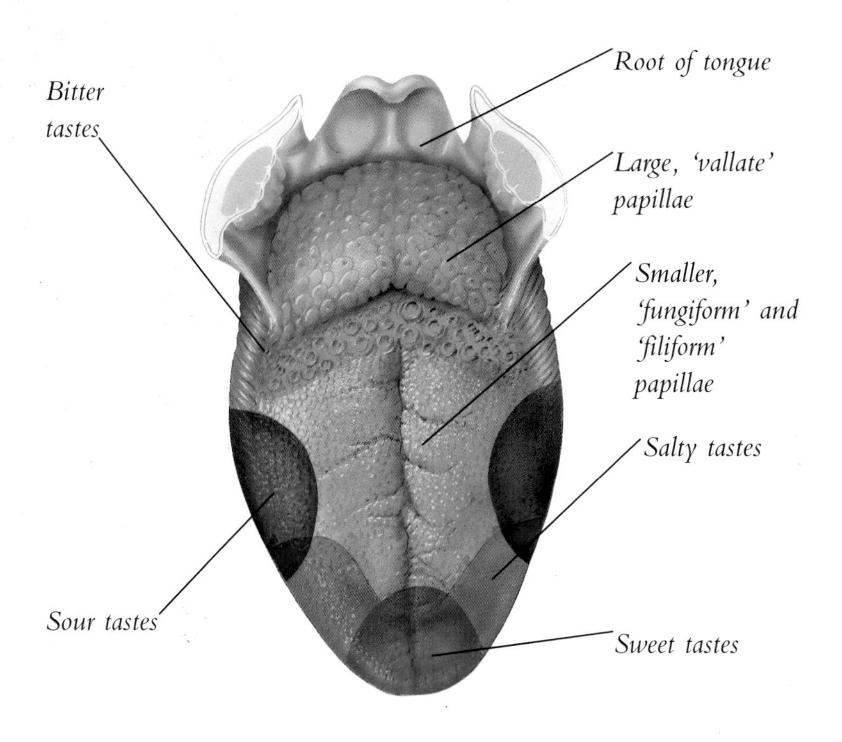

THE FOUR TASTES

The taste buds are located inside small bumps on the surface of the tongue called papillae. Taste buds that respond to sweet tastes are concentrated at the tip of the tongue. Salty flavours are detected most strongly just behind on the sides of the tongue. Sour tastes that make you wince produce the strongest response on the sides of the tongue, further back still. Really bitter tastes set off a cluster of taste buds right at the back of the tongue, just above the entrance to the throat.

Taste

- There are 10,000 or so taste buds in various places over the tongue.
- In children and adults, there are no taste buds in the centre of the tongue.
- A baby has taste buds all over the inside of its mouth, not just on its tongue.
- As you get older and older, your taste buds die off, and by the time you are 70 your sense of taste will be much less sensitive.
- Taste bud cells only last a week before the body renews them.
- You often lose your sense of taste when you get a cold because your nose is blocked and you can't smell – not because your taste buds have stopped working.
- Taste buds are not the only sense receptors in your mouth. There are also sensors for touch, pressure, moisture, heat, cold and various other factors.

NEW LIFE

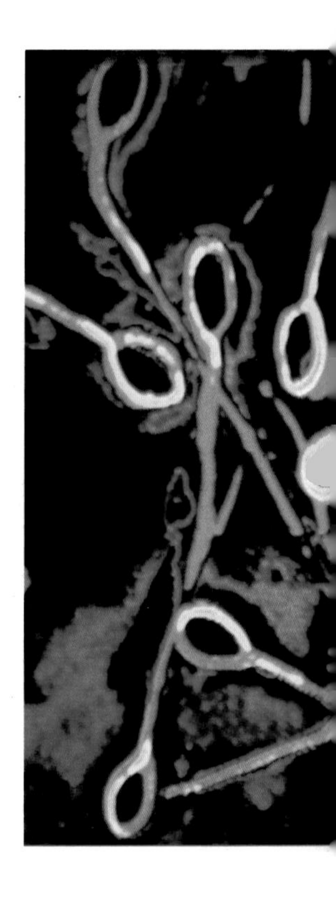

For a new human life to be created, a sperm from a man must fertilize an ovum or egg from a woman. Only when the 23 chromosomes in each are united to give the full complement of 46 will a baby begin to develop. Fertilization can now actually take place in a laboratory dish, but normally it occurs during sexual intercourse between a man and a woman.

At the climax of intercourse, millions of tiny sperm are ejaculated from the man's penis into the woman's vagina. The sperms then swim up the uterus towards the woman's Fallopian tube, where there may be an egg waiting. Very few sperm make it to the right tube, but only one is needed to fertilize the egg.

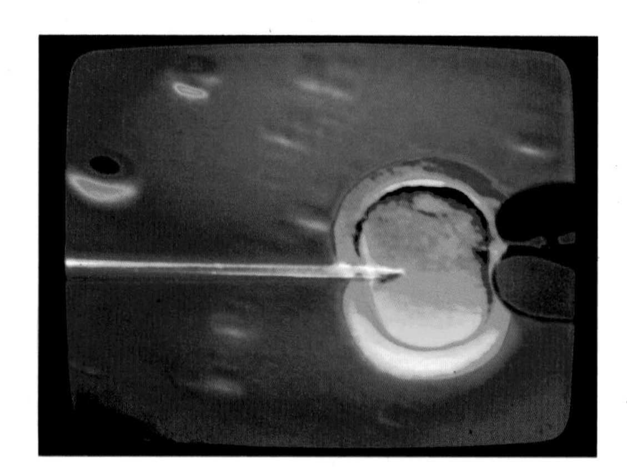

TEST TUBE FERTILIZATION
To fertilize an egg in a laboratory, an egg is taken from the potential mother's ovaries and joined with a sperm from the father. Once the egg is fertilized and starts to divide, it is usually returned to the mother's womb. This is called 'in vitro' fertilization because it happens in a glass dish — 'vitro' is the Latin for glass. In vitro fertilization can often lead to multiple births — twins, triplets and even sextuplets.

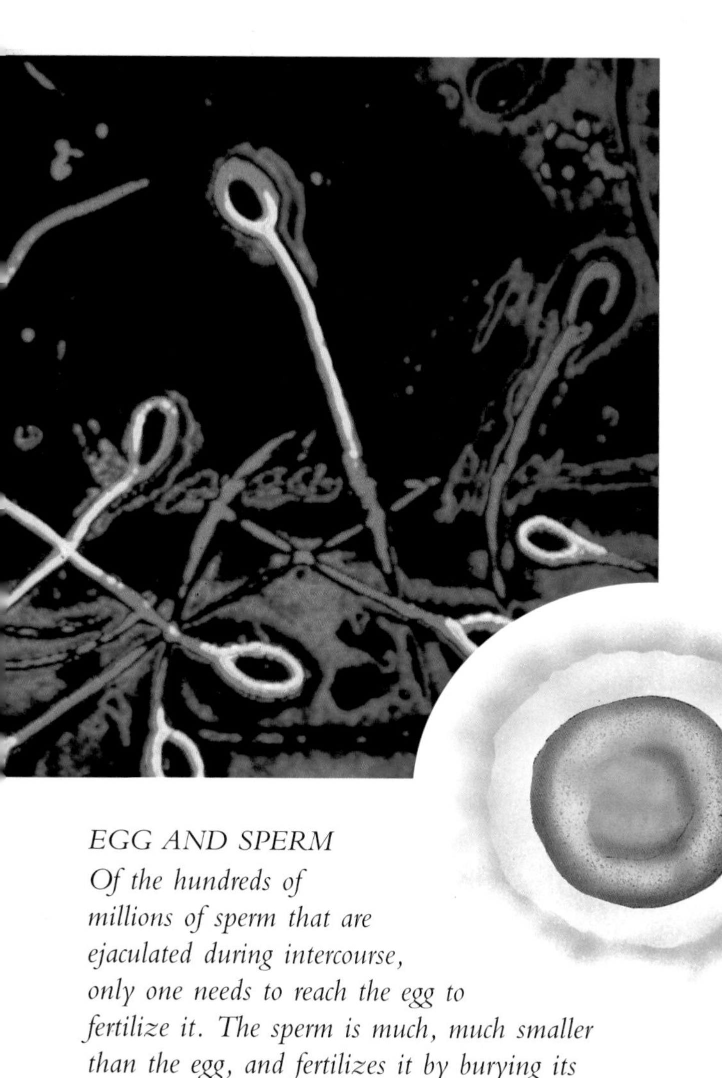

EGG AND SPERM
Of the hundreds of millions of sperm that are ejaculated during intercourse, only one needs to reach the egg to fertilize it. The sperm is much, much smaller than the egg, and fertilizes it by burying its head in the egg's huge side. As it is engulfed by the egg, its tail drops off and other sperm are shut out. Soon after, the egg divides, and prepares to unite with the sperm to form the first cells of the new life.

Sperm and egg race

- Sexual intercourse does not always result in a new life being conceived. The egg can only be fertilized in the 12 to 24 hours immediately after the woman ovulates.
- For the egg to be fertilized, a man's sperm must reach the woman's egg when it is moving down the Fallopian tube, halfway through the woman's menstrual cycle.
- When sperm are ejaculated during intercourse, they land near the woman's cervix, the neck of her uterus.
- Sperm look like tadpoles, and they swim up the uterus by waggling their tails.
- As the sperm approach the egg, their heads dissolve in chemicals released by the egg. But as they do, they send out an enzyme that helps one penetrate the egg. It takes the enzymes from hundreds of sperm to break down the egg's protective barrier — but one may finally make it through.

PREGNANCY

Pregnancy begins the moment an ovum or egg is fertilized by a sperm. The egg then divides inside the mother to make more cells and grows into an embryo, and after eight weeks, into a foetus. Unlike an embryo, a foetus has limbs and internal organs and it grows rapidly, protected inside its mother's uterus, now called the womb — and the womb grows with the foetus. After around nine months, the foetus is fully developed and ready to be born.

1. 4 weeks
Heart starts to beat

2. 5 weeks
Arm and leg buds appear

3. 8 weeks
Embryo becomes foetus as limbs form. Fingers and toes are webbed

4. 12 weeks
The head is now quite large, though the eyes are closed. Finger and toe nails start to grow

5. 16 weeks
The foetus is now recognizably a baby, covered in downy hair called lanugo. It may begin to kick

6. 20 weeks
If the foetus is a boy, its genitals may be visible. Lungs are formed, and the foetus can grip

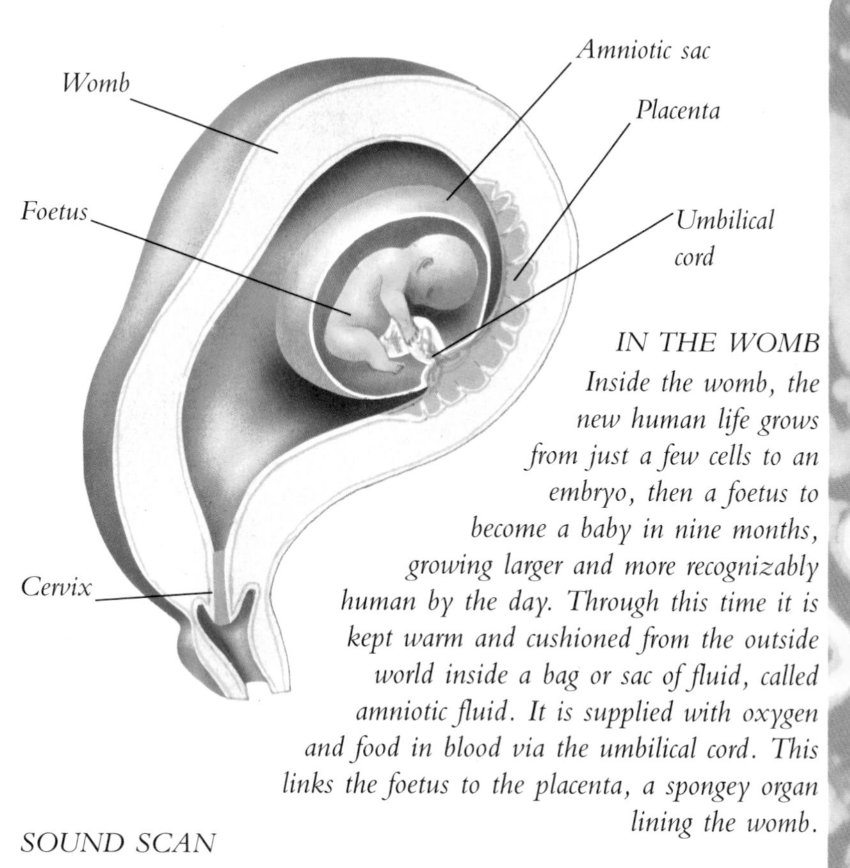

IN THE WOMB
Inside the womb, the new human life grows from just a few cells to an embryo, then a foetus to become a baby in nine months, growing larger and more recognizably human by the day. Through this time it is kept warm and cushioned from the outside world inside a bag or sac of fluid, called amniotic fluid. It is supplied with oxygen and food in blood via the umbilical cord. This links the foetus to the placenta, a spongey organ lining the womb.

SOUND SCAN
16 to 18 weeks into pregnancy, an expectant mother may go for an ultrasound scan. This uses sound waves too high for humans to hear to create a picture of the inside of the womb. The foetus can be seen moving on a display screen if it is healthy.

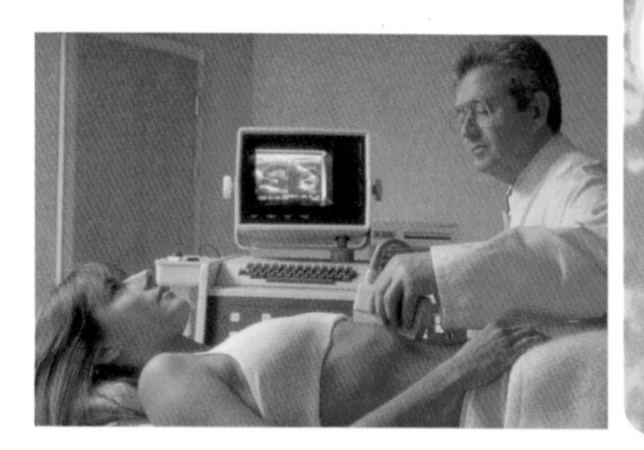

Pregnancy

- A woman can often tell that she is pregnant because her monthly period. A urine test confirms the pregnancy.
- Pregnancy is divided into three periods or 'trimesters': 0-12 weeks, 13-28 weeks and 29-40 weeks.
- In the first 6-8 weeks, a pregnant woman is often sick in the morning.

Body changes

- During pregnancy, a mother's body changes shape to meet its demands.
- She gains 30% more blood.
- Her heart rate goes up.
- She eats more and puts on weight. Surprisingly, less than half this extra weight is the foetus itself.
- Her belly begins to expand.
- The hormone oestrogen makes the breasts grow larger and develop milk glands.
- The hormone progesterone relaxes the muscles of the abdomen to allow the womb to expand.

INDEX

Acknowledgements

The publishers wish to thank the following artists who have contributed to this book.

Kuo Kang Chen, Andrew Clark, Jeremy Gower, Sally Launder, Mike Saunders, Guy Smith, Steve Weston.

All photographs from the Miles Kelly Archive.